SOIL FERTILITY AND PLANT NUTRIENT MANAGEMENT

About the Authors

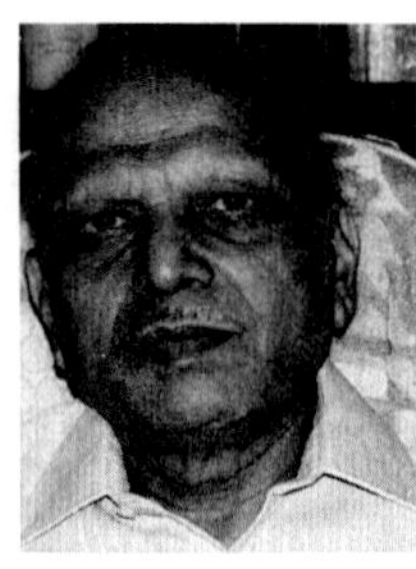

Professor (Dr) Rajendra Prasad was born on 1 January 1936 at Haidarabad, Unnao, Uttar Pradesh, India. Dr Prasad fetched his M.Sc. (Agriculture) (1956) degree from the Agra University, Agra, Uttar Pradesh, and Ph.D. degree (1961) from the Mississippi State University, Starkville, Mississippi, USA. He is a scientist, teacher, administrator and author of over 55 years of experience in the field of Agronomy and provided services to the nation independently and through his students. Dr Prasad has published over 300 research papers in national and international journals of repute and also participated in a number of national and international seminars/conferences/ congresses etc.

Prof Prasad worked in various capacities like Assistant Agronomist, Mississippi State University (1961), Assistant Professor of Agronomy (1963–64), Agronomist (1964–73), Professor of Agronomy (1973–83 and 1986–91), Head, Division of Agronomy (1983–86 and 1991–95), ICAR National Professor (1996–2000), and INSA Senior Scientist (2004–05), ICAR–Indian Agricultural Research Institute, New Delhi. He was also visiting Scientist, University of Nebraska, Lincoln, USA, 1993; and UNESCO Consultant to SR Vietnam (1986, 1987).

Prof Prasad is the recipient of numerous national awards including Hooker Award of the ICAR–Indian Agricultural Research Institute (New Delhi), the Rafi Ahmed Kidwai Award of the Indian Council of Agricultural Research (New Delhi), and the Silver Jubilee Award of the Fertiliser Association of India (New Delhi) for his outstanding contributions in soil fertility and fertilizer nitrogen management. Prof Prasad was bestowed with the Indian Society of Agronomy Gold Medal 1995 and also Lifetime Achievement Award by the Indian Society of Agronomy, 2012. He is a Fellow of the Indian National Science Academy and National Academy of Agricultural Sciences (India). Dr Prasad also served the Indian Society of Agronomy (1975–78) as Secretary and as Vice President and Executive Chairman (1979–80). He was Editor-in-Chief, Indian Journal of Agronomy during 2011–16 and a member of Editorial Board of the Journal of Agronomy and Crop Science, Berlin. Dr Prasad has taught a course on soil fertility and its management at the ICAR–IARI for over 30 years and has guided over 32 students for Ph.D. degrees in agronomy. Dr Prasad grew up in India and now live with his family in Buffalo, New York, USA.

Dr Y.S. Shivay was born on 1 December 1966 at Fuldehra, Uttar Pradesh. He received his education at the Chandra Shekhar Azad University of Agriculture and Technology, Kanpur and the Govind Ballabh Pant University of Agriculture and Technology, Pantnagar, for his B.Sc. (Ag. & AH) (1986), M.Sc. (1989) and Ph.D. (1995) degrees. Dr Shivay is an excellent scientist and teacher, having experience of more than 29 years in the field of Agronomy. He is a skilled writer too. He has worked as Scientist (1991–2000), Senior Scientist (2000–08) and Principal Scientist (2008 to date), at the ICAR–Indian Agricultural Research Institute, New Delhi, India. Dr Shivay also served as Professor, at the Division of Agronomy of this institute during 2015 to 2018. He was nominated visiting Scientist at the Zhejiang University, PR China during 1998–99, and also at The Norwegian University of Life Sciences, Aas, Norway during 2003–04 and 2008.

Dr Shivay has been bestowed with many national and international awards/ honours, viz. Chinese Government Scholarship 1998–99 through Ministry of Human Resource Development, Government of India; Norwegian Government Scholarship 2003–04 through Ministry of Human Resource Development, Government of India; Distinguished Scientist Award 2009; ICAR–IARI Best Teacher Award 2009; FAI Golden Jubilee Award 2012; FAI Shriram Puraskar 2012, 2015, 2017; SAP Gold Medal 2015; IZA–FAI Award 2016. He was elected as Organizing Secretary for the 4th International Agronomy Congress 2016. Dr Shivay received Letter of Appreciation for Imparting the Teaching and Research Guidance to the ANASTU Students from Islamic Republic of Afghanistan 2016, 2018 and 2019. He also received The Falvey Award 2016, ICAR Bharat Ratna Dr C. Subramaniam Award for Outstanding Teachers 2018 and Best Poster Award XIV Agricultural Science Congress.

Dr Shivay is a Fellow of the Indian Society of Agronomy, Indian Society of Plant Physiology and National Academy of Agricultural Sciences (India). He also served as the Secretary of the Indian Society of Agronomy during 2015–19 and as treasurer for 1995–96. Dr Shivay has over 26 years experience of teaching on soil fertility and its management at the ICAR–IARI, New Delhi and has guided more than 17 students for obtaining their M.Sc. and Ph.D. degrees in Agronomy. His students are at senior positions in the Indian Council of Agricultural Research and Central Agricultural Universities in India. Students, teachers and scientists are always in demand for his published work for updating their knowledge.

SOIL FERTILITY AND PLANT NUTRIENT MANAGEMENT

Rajendra Prasad
PhD (Mississippi State) FNA, FNASc, FNAAS, FISA
Ex-Professor and Head, Division of Agronomy
and Ex-ICAR National Professor
ICAR–Indian Agricultural Research Institute
New Delhi 110 012

Yashbir Singh Shivay
PhD (GBPUAT, Pantnagar), FNAAS, FISA, FISSP
Principal Scientist and Ex-Professor of Agronomy
Division of Agronomy
ICAR–Indian Agricultural Research Institute
New Delhi 110 012

NEW INDIA PUBLISHING AGENCY
New Delhi 110 034

NEW INDIA PUBLISHING AGENCY
101, Vikas Surya Plaza, CU Block, LSC Market
Pitam Pura, New Delhi – 110 034, India
Email: info@nipabooks.com
Web: www.nipabooks.com
For customer assistance, please contact
Phone: + 91-11-27 34 17 17 Fax: + 91-11- 27 34 16 16
E-Mail: feedbacks@nipabooks.com

ISBN : 978-93-90512-27-0

Composed and Designed by NIPA

Dedication

Dedicated to Graduate Students and Alumni in
Agronomy, Soil Science and Agricultural Chemistry,
Agricultural Physics and Plant Physiology at the
ICAR–Indian Agricultural Research Institute
New Delhi

त्रिलोचन महापात्र, पीएच.डी.
सचिव, एवं महानिदेशक

TRILOCHAN MOHAPATRA, Ph.D.
SECRETARY & DIRECTOR GENERAL

भारत सरकार
कृषि अनुसंधान और शिक्षा विभाग एवं
भारतीय कृषि अनुसंधान परिषद
कृषि एवं किसान कल्याण मंत्रालय, कृषि भवन, नई दिल्ली 110 001

GOVERNMENT OF INDIA
DEPARTMENT OF AGRICULTURAL RESEARCH & EDUCATION
AND
INDIAN COUNCIL OF AGRICULTURAL RESEARCH
MINISTRY OF AGRICULTURE AND FARMERS WELFARE
KRISHI BHAVAN, NEW DELHI 110 001
Tel.: 23382629; 23386711 Fax: 91-11-23384773
E-mail: dg.icar@nic.in

Foreword

With the advent of high yielding varieties of wheat and rice and the hybrids of maize, sorghum and millets in India, fertilizer application has considerably increased and has been the key input in increasing crop yields and attaining self-sufficiency in food production in the country. However, the efficiency of plant nutrients is very low in India as compared to that in advanced countries, because these are not adequately and properly applied. Therefore, a better understanding of the basic principles of Soil Fertility and efficient use of plant nutrients is important. A knowledge of reactions that various plant nutrients undergo in soils and their deficiency symptoms have become crucial for the students of agriculture, especially those majoring in Agronomy, Soil Science and Plant Physiology, so that they can appropriately educate the agricultural extension workers and the farmers. A number of foreign books on the subject are available but their emphasis is more on soil and environmental conditions of other countries, hence a book dealing with Indian conditions is very much needed. The present book discusses the fundamentals of soil fertility conditions in India and the reactions that various plant nutrients undergo in Indian environmental conditions and fulfill the plant need.

I am happy that the book entitled 'Soil Fertility and Plant Nutrient Management' has been authored by an experienced and renowned agronomist, Prof. Rajendra Prasad, Ex ICAR National Professor and Dr. Y.S. Shivay, Principal Scientist & Ex Professor, IARI. I am confident that teachers, researchers, extension scientists and students will find this book very useful.

Dated the 10th August, 2020
New Delhi

(T. MOHAPATRA)

Preface

This book has been written to cover the contents of the Post-Graduate course on 'Soil Fertility and Nutrient Management' offered by the departments of Agronomy and Soil Science at the state agricultural universities in India and at the ICAR–Indian Agricultural Research Institute, New Delhi. The narrative in the book is brief and to the point in a simple and easy to understand language, demanding least possible time of students. A few important references are provided in each chapter for further reading. Also at the end of each chapter a few questions of varying kind are provided to recapitulate the main points. This book is intended for preparing the Post-graduate students for their university examinations, competitive examinations and interviews. It is short and permits revision in a short time.

Although written for Post-Graduate students in Agronomy and Soil Science in India, this book should be equally useful to students in neighbouring countries and to Agronomists and Soil Scientists in the Fertilizer industry in the country as a handy guide book for plant nutrient management. We have made an attempt to include some excellent data generated by the Agronomists and Soil Scientists in India and neighbouring countries, while discussing different topics and an author index is provided at the end. If a researcher/teacher in Agronomy and Soil Science feels that he has generated some good data on an important topic in soil fertility and plant nutrient management he should send a soft or hard copy of a reprint of that paper to us. We will try our best to include their research findings in future editions of this book. Comments from the teachers, researchers and students are requested.

Rajendra Prasad
Yashbir Singh Shivay

Contents

Chapter 1

Soils as the Medium for Crop Production

LAND VERSUS SOIL

Land is a part of Earth's surface that is not covered by water. It is used for agriculture, building houses, office buildings, schools, hospitals, roads, railroads and even water canals. Land is measured by area [length (l) × width (w)] and the units are square meters (m^2), square centimeters (cm^2) etc. On the other hand, soil is Earth's crust, which could vary from a few centimeters to a few meters in thickness (depth), needs a measure of volume [length (l) × width (w) × depth (d)] and the units are cubic meters (m^3), cubic centimeters (cm^3) etc. Soil is thus a three-dimensional natural body on Earth's surface that provides anchorage to growing plants/ trees and shelter to soil biota and is responsible for the storage of water and plant nutrients, which are essential for crop production. In addition, it helps in the biodegradation of soil pollutants and purification of ground water, the major source of drinking water in the world. Soil feeds all living beings on Earth and its degradation is one of the reasons for inadequate human nutrition (Lal, 2009).

SOIL AS A LIVING BODY

Soil is made up of all the three phases of matter, namely, solid, liquid (water) and air. In general, it is 50% solid and 50% (water + air); the proportion of air and water keeps on changing depending on the amount of rain water received or irrigation water applied. The solid phase has three components, namely, mineral matter, organic matter (plant and animal residues at different stages of decomposition) and soil biota including microorganisms (bacteria, fungi, actinomycetes etc.), meso fauna (mites, collembolan etc.) and macro fauna (mites, ants, earthworms etc.). Air and water occupy the empty space between the solid mineral and organic matter particles, which is known as "pore space".

Most upland crops, viz. wheat (*Triticum aestivum* L.), barley (*Hordeum vulgare* L.), maize (*Zea mays* L.), sorghum [*Sorghum bicolor* (L.) Moench], chickpea (*Cicer arietinum* L.), mungbean or greengram [*Vigna radiata* (R) Wilczek], pigeonpea [*Cajanus cajan* (L.) Millsp.], cotton (*Gossypium* sp.), sugarcane

(*Saccharum officinarum* L.), require adequate aeration, which is essential for root respiration and plant growth. Roots absorb oxygen for respiration that releases energy which is utilized for the transportation of nutrients from the soil. Water logging adversely affects the growth of most crops, except rice (*Oryza sativa* L.). However, some crops, such as maize (Prasad *et al.*, 2004), soybean [*Glycine max* (L.) Merr.] (Sorte *et al.*, 1995) and pulses (Krishnamoorthy *et al.*, 1987; Kumutha *et al.*, 2009; Kumar *et al.*, 2013) are very sensitive to water logging. On the other hand, rice (*Oryza sativa* L.) can grow under waterlogged conditions, when the air concentration in pore space is near zero. This is because, rice plant has aerenchyma tissue in the leaves, stem and roots, which allows it to get the oxygen for root respiration from the air in the atmosphere (Keddy, 2010).

Because of pore space, the soil has two densities, one with the air space as it occurs in nature and is known as bulk density (BD). The other is the particle density (PD) or true density, based on exclusion of the volume occupied by air and water. The unit for both kinds of soil densities is same and is g/cubic centimeter (g cc^{-1}). For a method of determining bulk density of soil, reference may be made to a book on Soil Physics (e.g. Shukla, 2013). The formula for determining percentage pore space is as under:

Pore space (%) = 100 (1–BD/PD),

For example, if BD is 1.3 g cc^{-1} and PD is 2.6 g cc^{-1}, the pore space (%) is 100 (1–1.3/2.6) = 50%

MINERAL MATTER

Soil mineral matter consists of particles of different sizes. As per the International Society of Soil Science, these particles and their sizes (diameter) are: gravel 2 mm or more; coarse sand 0.2–2 mm; fine sand 0.02–0.20 mm; silt 0.002–0.02 mm; clay $\leq$ 0.002 mm. The percentage of sand, silt and clay present in a soil gives it a *soil textural class* name, e.g. sandy loam (70–85% sand, < 15% clay and the rest as silt); loam (< 50% sand, 10–30% clay and the rest as silt); silt loam (70–90% silt, 20–50% sand and the rest as clay). Soil textural class can be found out after determining the percentage of sand, silt and clay and using a triangular diagram as given in Fig.1.1. Soil textural class is an important characteristic of soil and determines its water- and nutrient-holding capacity, as well as its maneuverability for cultivation.

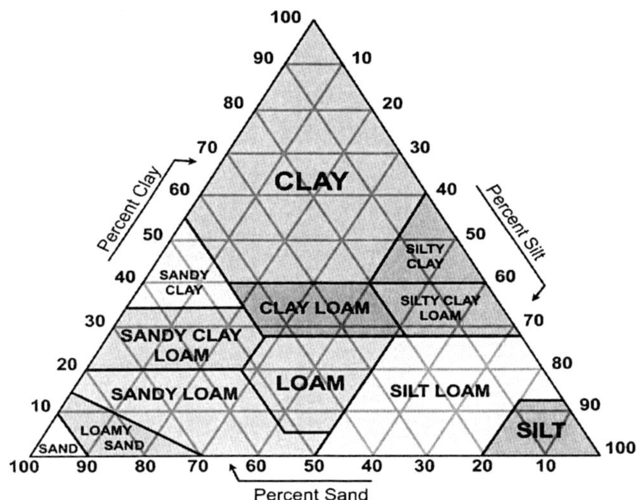

Fig.1.1: Chart for determining soil textural class (Soil Survey Staff, USDA, 1999)

Another soil property related to soil particles is "soil structure". While soil textural class tells the per cent distribution of different size particles, soil structure describes how these particles are held together. The important structure types are described below:

Granular: Soil particles held together by soil organic matter (SOM) in small granules, easily breakable by fingers, such as in alluvial loam soils. This is an ideal structure for most upland crops. These soils are easy to manage.

Blocky: Soil particles held together by clay particles and some SOM in small blocks (bigger than granules) not easy to break by hand.

Columnar: Soil particles held together by the clay particles and some SOM forming a column, such as in black cotton soil on drying.

Sometime columns are prismatic in shape, giving a definite prismatic structure.

Platy: Soil with high clay content, tends to form plate like structures and these plates sit over each other making the soils slippery. The examples are soils at the bottom of a pond or lake. Many rice areas have such soil.

Clay particles

Clay particles, because of their very small size ($\leq$0.002 mm) possess colloidal properties. Clay particles are mostly made of clay minerals, which are layered silicates made of layers of Si-O/OH tetrahedra and Al-O/OH octahedra. Tetrahedra involve three O or OH at the base and one O/OH on the top and a Si atom in the centre, while an octahedral layer involves three O/OH on the top as well as at the bottom and an Al in the centre; the O/OH of the tetrahedral and octahedral layer are shared. For all practical purpose one is really concerned with 2 groups of clay minerals, namely, 1 : 1 layer silicates (1 tetrahedral : 1 octahedral layer) [the kaolinite/halloysite group] and 2 : 1 layer silicates (2 tetrahedral : 1 octahedral layer [the pyrophyllite/mica/smectite group]. The basic ideal unit formulae (having all the properties of a mineral) for some clay minerals are given below. The negative charge on mineral particles is expressed as centimoles per kilogram [c (+) mol kg^{-1}].

Kaolinite	$Al_2^{3+}\ Si_2^{4+}\ O_5^{-2}\ (OH)_4^{-1}$ (unit charge 0) [net negative charge 2–5 c (+) mol kg^{-1}]
Pyrophyllite	$Al_2^{3+}\ Si_4^{4+}\ O_{10}^{-2}\ (OH)_2^{-1}$ (unit charge 0) [net negative charge 0 c (+) mol kg^{-1}]
Mica (Muscovite)	$Al_2^{3+}\ (Si_3^{4+}Al^{3+})^{-1}\ O_2^{-2}\ (OH)_4^{-1}$ (unit charge–1) (Net negative charge 0 c (+) mol kg^{-1} due to adsorption of K^+ +or NH_4^+ or other cations and closure of the lattice)
Smectite (Montmorillonite)	$(Al_{1.7}^{3+}\ Mg_{0.3})^{-0.3}\ (Si_{3.9}^{4+}Al_{0.1}^{3+})^{-0.1}\ O_2^{-2}\ (OH)_4^{-1}$ (unit charge–0.4) [Net negative charge 80–120 c (+) mol kg^{-1}]

Net negative charge and its importance in soil fertility

Silicon in the tetraderal layer and Al in the octahedral layers provide a completely balanced structure as in Kaolinite and pyrophyllite. However, in nature, while these minerals are formed or weather, some impurities creep in and a part of Si (54 pm ionic radius) may be substituted by a nearly same size element Al (67.5 pm ionic radius) (Table 1.1) in pyrophyllite and this leads to the formation of another group of minerals, namely, micas. This substitution of elements is known as isomorphous substitution, which means substitution of one element by another element of nearly the same size. However, when Al^{3+} substitute Si^{4+}, it creates a –1 mole (negative) charge on the mineral particle and the mineral can attract cations (K^+, NH_4^+, Ca^{2+}, Mg^{2+}, Zn^{2+}) etc. on its surface. Similarly, Al^{3+} in the octahedral layer could be replaced by Mg^{2+} and create a negative charge on the surface of the clay mineral. Thus, isomorphous substitution creates negative charge on the surface of the clay mineral, which makes it possible to hold the cations on its surface. These cations can be exchanged with other cations in soil solution and this is why the surface layer of clay mineral is referred to as exchange complex. However, in general, there is partial isomorphous substitution in most cases and this is why the negative charge created is generally a fraction of a mole, e.g. –0.3 in the octahedral layer and –0.1 in tetrahedral the layer of montmorillonite (a smectite group mineral) in the unit formula given above. There are several members in the smectite group, with different degrees of isomorphous substitution and the resultant charge. Even the ions involved in isomorphous substitution differ and this is what gives different names to different minerals (for details, reference may be made to Prasad and Power, 1997; Barton and Karathanasis, 2002). The negative charge on minerals contributes to its cation-exchange capacity (CEC), which is expressed as c (+) moles kg^{-1} mineral/soil. The earlier unit was me (milli equalent) 100 g^{-1} mineral/soil and numerically has the same value as c (+) moles kg^{-1} mineral/soil; c-mole stands for centimoles and (+) is for positive ions or cations.

Table 1.1: Properties of some elements involved in the structure of clay minerals

Element (Symbol)	Atomic weight	Valence	Atomic radius (pm)	Ionic radius (pm)
Aluminum (Al)	26.981	+3	118	67.5
Calcium (Ca)	40.078	+2	194	114
Hydrogen (H)	1.008	+1	53	2
Iron (Fe)	55.845	+3	156	72.5
Magnesium (Mg)	24.305	+2	145	86
Oxygen (O)	15.999	–2	48	126
Potassium (K)	39.098	+1	243	143–152
Silicon (Si)	28.085	+4	111	54

pm (Picometer) = 10^{-12} m = 10^{-9} mm = 10^{-2} A (Angstrom); 1 Angstrom = 100 pm

Note: In general, ionic radii of cations are generally lesser than atomic radii, while the ionic radii of anions are larger than their atomic radii. Ionic radii also vary depending on the valence change in a reaction.

Source: Rao (1999)

In the micas, e.g. in Muscovite, the Al-mica, there is –1 mole negative charge, but it is completely balanced by K^{+1}, which fits very well in the void left between 6-oxygen atoms between the Si- tetrahedral and Al-octahedral layers and gets fixed and is not available for exchange. This is known as potassium fixation. The NH_4^{+1} has almost same size as K^{+1} and gets fixed in the micas same way. Micas other than Muscovite are Phlogopite (Mg mica) formed by isomorphous substitution of Si^{4+} by Mg^{2+} and Biotite (Fe mica) formed by isomorphous substitution of Si^{4+} by Fe^{2+} or a combination of Mg^{+2} and Fe^{+2}. Thus, micas are very rich in K and this is the reason why soils in the Ganga-Yamuna region are so rich in K, because the silt brought in by the rivers is rich in micas brought in from the Himalayas. The fixed K in north Indian alluvial soils could range from 0.79 to 0.90 g K kg^{-1} soil (Grewal and Kanwar, 1976).

Calculating cation exchange capacity of clay minerals

Since the negative charge on clay minerals is created by isomorphous substitution, it is generally not present in 1 : 1 layer silicates. The little charge [2–5 c (+) moles kg^{-1} soil] kaolinite has is due to OH^{-1} on broken edges of the mineral particles.

The calculation of negative charge on 2 : 1 clay minerals is explained here using the basic unit formula Al_2^{3+} Si_4^{4+} O_{10}^{-2} $(OH)_2^{-1}$. The formula weight of this basic unit is 360 (g). If this unit acquires a charge of –1 mole it is equal to 2.78 moles (+) moles kg^{-1} soil (1,000/360=2.78). Thus, a 0.4 mole charge on mineral montmorillonite equals 1.11 m (+) moles kg^{-1} montmorillonite or 111 c (centimoles) (+) moles kg^{-1} montmorillonite. This is known as cation-exchange capacity (CEC) of a mineral. In general, the CEC of 2 : 1 minerals is about 80–120 c (+) moles kg^{-1}. As a common practice, the CEC is determined for the soil as a whole involving all minerals and other particles. The CEC in silicate minerals is due to structural isomorphous substitution and is permanent in nature and is not dependent on soil *p*H. The CEC of some minerals is given in Table 1.2.

In Ultisols and Oxisols which are found in several tropical and subtropical countries, most of the Si is leached out and silicate minerals are not present in abundance. In such soils, Fe and Al oxides and hydroxides dominate, giving them the names, such as, red soils, yellow soils. In India such soils are found in southern and north-eastern states. In these soils the CEC is exhibited by the oxides and hydroxides of iron and is *p*H dependent, and such charge is known as variable charge

Table 1.2: Cation exchange capacity (CEC) of some clay minerals

Clay mineral	Ideal formula	Charge per unit formula	Net negative charge [c(+) mol kg^{-1})]
1:1 clay minerals			
Kaolinite	$AlSi_2O_5(OH)_4$	0	2–5
Antigorite	$Mg_3\ Si_2\ O_5(OH)_4$	0	NA
2:2 clay minerals			
Muscovite mica	$Al_2(Si_3\ Al)^{-1}O_{10}\ (OH)_2$	–1 (charge satisfied by 1.0 K^+)	0
Illite	$Al_2(Si_{3.2}Al_{0.8})^{-0.8}O_{10}\ (OH)_2$	–0.8	15–40
Vermiculite	$(Al_{1.7}\ Mg_{0.3})^{-0.3}(Si_{3.6}Al_{0.4})^{-0.4}O_{10}\ (OH)_2$	–0.7	100–180
	or		
	$(Mg_{2.7}Fe^{3+}_{0.3})^{+0.3}(Si_3Al)^{-1.0}O_{10}\ (OH)_2$	–0.7	100–180
Montmorillonite	$(Al_{1.7}\ Mg_{0.3})^{-0.3}(Si_{3.9}Al_{0.4})^{-0.1}O_{10}\ (OH)_2$	–0.4	80–120
Chlorite	$(Mg_{2.6}\ Fe^{3+}_{0.4})^{+0.4}(Si_{2.5}Al/Fe_{1.5})^{-1.5}O_{10}\ (OH)_2$	–1.1 (charge satisfied by $Mg_2Al\ (OH)_6^{+1}$ layer	15–40

Source: Prasad and Power (1997)

(Prasad and Power, 1997; Qafoku *et al.*, 2004). Only under alkaline conditions these soils acquire CEC and therefore, in general, are poor in fertility. Under acid conditions, which generally prevail in such soils, they acquire a positive charge and exhibit anion-exchange capacity. They can absorb large amounts of OH^- ions and require large amounts of liming material to neutralize their acidity (Sanchez, 1976). Also they need larger application of phosphate (PO_4^{3-}) fertilizer to overcome P deficiency. To prevent rapid fixation of applied water-soluble P, application of ground phosphate rock is recommended for such soils.

In addition to silicate and Fe and Al oxide and hydroxide minerals, soil organic matter, because of the presence of [$COOH^-$] and [OH^-] radicals also exhibits considerable CEC (60–120 c (+) moles kg^{-1}) or even more depending on its composition.

Use of cation-exchange capacity in soil testing and crop production

An example of the knowledge of CEC in soil testing is its use in determining available (exchangeable) K in soil. For this purpose a 5 g soil sample is taken in a 100 ml Erlenmeyer flask and 25 ml of the 1 N ammonium acetate solution is added and the contents in the Erlenmeyer flask are shaken for 5 minutes. This permits the exchange of K in the soil-exchange complex with NH_4^+ in the solution. The exchanged K in solution is estimated and is known as available (exchangeable) K in soil (Prasad *et al.*, 2013).

Another example of using the knowledge of CEC in crop production is the reclamation of sodic (alkali) soils by the use of gypsum ($CaS_4.2H_2O$). Once gypsum is applied and incorporated in soil and the soil is repeatedly flooded, over a period of time (few months to few years), Ca in the gypsum replaces Na in the exchange complex of soil clay minerals, which is then leached down the profile and increased Ca-saturation of the exchange complex makes the soil suitable for crop production (Singh and Abrol, 1988).

SOIL ORGANIC MATTER

Soil organic matter (SOM) includes components varying from partially decomposed plant and animal residues to the most important completely decomposed dark humus. It is humus, which is generally referred to as soil organic matter. The SOM serves as a substance for soil microbial activities, provides binding glue to soil particles and helps improve soil structure, stores moisture and nutrients and supplies these to the plants. Plant nutrients are held as a constituent of SOM and are released on its decomposition. This is important for the release of N and S, which are components of protein residues, P from phospholipid and other P-containing residues and other plant nutrients. On decomposition SOM also releases K and other nutrients. The SOM also has CEC due to the presence of phenolic ($-C_6H_5OH$) and carboxylic ($-COOH$) groups and it can vary from 10 to 30 c (+) mole kg^{-1} SOM. No wonder SOM has been referred to as the elixir of all terrestrial life (Lal, 2014). The relationship between SOM and different soil-quality parameters are shown in Table 1.3.

Table 1.3: Relationship between soil organic carbon (SOC) with various indicators of soil quality

Soil quality parameters	Soil type	r^*
Hydraulic conductivity	Inceptisol	0.82
Porosity	Alfisol	0.72
Available N	Entisol, Inceptisol, Alfisol	0.93
Soil microbial biomass carbon (SMBC)	Inceptisol	0.90
SMBC	Inceptisol	0.92
Dehydrogenase	Alfisol	0.81
Urease	Alfisol	0.94

*Statistically signifcant at 1% level

Source: Rao *et al.* (2019)

Soil organic matter in Indian soils can vary from <0.02% in desert soils in dry hot regions of Rajasthan to 3% or even more in hill soils in cold hilly regions of Himachal Pradesh and Uttarakhand.

SOIL BIOTA

Soil biota includes microorganisms and soil micro fauna (body width < 0.01 mm), mesofauna (body width 0.01–2.0 mm) and macrofauna (body width > 2 mm) (Saxena *et al.*, 2014). A few details about these are as under. As regards micro-organisms, 1 g of soil from field of rice-based cropping systems at New Delhi may contain 6.1×10^6 cfu (colony forming units) of bacteria, 3.2×10^4 cfu fungi and 18.2×10^4 cfu actinomycetes (Sharma *et al.*, 2010).

Soil microorganisms

From the viewpoint of crop production bacteria, fungi are the most important microorganisms (Subba Rao, 1999).

Bacteria (about 500,000 species): Includes N-fixing rhizobia, *Azotobacter*, *Azospirillum*, ammonifiers, nitrifiers, denitrifiers, phosphate solubilizers, methanogens, methanotrophs, S-oxidizing bacteria, sulphate-reducing bacteria etc.

Non-mycorrhizal fungi (about 35,000 species): Decomposition of lignin, chitin, agrochemicals, mineralization of plant nutrients etc.

Mycorrhizal fungi– Ectomyccorhiza, vesicular arbuscular mycorrhiza (VAM)– (about 10,000 species): Mobilize water and nutrients (especially N) from soil to plant roots.

Soil Microfauna

Protozoa (about 100,000 species): Grazers of bacteria and fungi, high-quality feed to meso- and microfaunal grazers, parasitic to higher level of organisms.

Nematodes (about 500,000 species): Enhance C and N availability to plants, parasitic to several plant species, especially in vegetable and fruit crops.

Soil mesofauna

Mites (about 30,000 species): Grazers of bacteria and fungi, consume litter, roots and animal carcasses, feed on microfauna.

Insects – root herbivores (about 40,000 species): Grazing on above-ground herbivore populations, can cause reduction in crop yields.

Collembola (6,500 species): Grazing of microflora, consumers of plant litters and animal carcasses.

Ants: Ecosystem engineers, transporters of SOM, makers of holes and hills, stores of food.

Enchytreids: Consume plant residues, pests on agricultural crops.

Soil macrofauna

Earthworms (about 3,000 species): Ecosystem engineers, producers of castes of different sizes and shapes, mixing of lower and upper layer soils, improve soil aeration, help in soil aggregation. There are 3 functional groups: (i) Epigeics – feeding on fresh or partially decomposed litter on soil surface and rarely making burrows; (ii) Endogeics – feeding on mineral soils in deeper horizons and rarely making permanent burrows and (iii) Anecics – making permanent burrows in deeper soils, keep coming to the surface and dragging residues to deeper layers (Saxena *et al*., 2014). Earthworms help in maintaining soil fertility and in rehabilitating degraded ecosystems. In recent years, there has been a great interest in vermicompost, which is prepared using earthworms (Misra *et al*., 2003).

Termites (about 2,000 species): Ecosystem engineers/ bioturbulators, makers of mounds as high as 8–10 m, consumers of woody residues.

MAJOR SOIL GROUPS OF INDIA

As per latest information from the Soil Survey Organization of India, 11 major soil groups/ classes occur in India (Table 1.4) (Bhattacharya *et al*., 2013). Of these, five groups cover 87.3% of the country's area. These are: alluvial (30.4%), red (26.8%), black (16.6), desert (8.0%) and laterites (5.5%). Names of soil groups/ class in India indicate the climatic and soil-forming processes responsible for the formation of different soils.

The USDA soil taxonomic classification reflects the dominant soil-forming factors active during soil formation at a particular location. The USDA system of Soil Taxonomy (soil naming) consists of a hierarchy of six levels, which are: order, suborder, great group, subgroup and family (Soil Survey Staff, 1999). An Indian soil group/ class may contain soils belonging to different orders. Special characteristics of the soil orders mentioned in Table 1.4 are given below:

Entisols: Little soil profile development as in the case of alluvial soils or desert soils.

Inceptisols: Soils in the beginning of soil formation, more profile development than in Entisols.

Alfisols: Well-developed soil profile with an argillic (clayey) subsoil.

Aridisols: Dry desert soils.

Mollisols: Soils with the top layer rich in organic matter, such as foothill on tarai and hill soils.

Ultisols: Intensely weathered soils of the tropics, rich in iron and aluminum, such as, red and lateritic soils.

Vertisols: Deep soils with high content of smectite (2 : 1) clay minerals. These soils swell when wet and shrink when dry. They develop deep cracks on drying, found in black soils.

SUMMARY

1. Soil is a three-dimensional body in contrast to land which is bi-dimensional.
2. The solid component of soil is made-up of inorganic and organic matter.
3. Soil mineral matter has three kinds of particles, namely, sand, silt and clay.
4. Clay particles form the colloidal complex and are made of silicate minerals. There are two major kind of minerals, 1 : 1 layer silicates, such as Kaolinite and 2 : 1 kind of minerals, such as smectites. Cation-exchange capacity of 2 : 1 minerals is about 80–120 c (+) moles kg^{-1}, while that of 1 : 1 layer silicates is 5–10 c (+) moles kg^{-1}. Due to CEC, clay minerals hold most cationic nutrients (Ca, Mg, Fe, Zn, Cu, Mn etc) in their exchange complex.

Table 1.4: Extent and distribution of the different soil groups/classes of India along with their soil orders as per USDA nomenclature system

Major soil group	Area (M ha)	%	States, where found	Soil order (USDA classification)
Alluvial, Entisols	100	30.4	J&K, HP, Pb, Har, Del, UP, Guj, Goa, MP, Mah, AP, Kar, TN, Bihar, Od, WB, AP, Asom, Nag, Man, Miz, Tri, A&N	Inceptisols, Alfisols, Aridisols
Coastal alluvial	10	3.1	AP, Kar, TN, Ker, WB, Guj, Od, Pod, A&N	Aridisols, Inceptisols
Red	88	26.8	Kar, Ker, TN, Pod, Raj, MP, Mah, Guj, Goa, ArP, Assam, Man, Megh, Nag, Tri, Del, HP, A&N	Alfisols, Ultisols, Entisols, Inceptisols, Mollisols, Aridisols
Laterites	18	5.5	AP, Kar, Ker, TN, Pod, Mah, Od, WB,	Alfisols, Ultisols, Inceptisols
Black	54.7	16.6	MP, Mah, Raj, Pod, TN, Bihar, Od, Guj	Vertisols, Mollisols, Inceptisol,
Desert	26.3	8.0	Raj, Guj, Har, Pb	Entisols, Aridisols, Inceptisols, Entisols
Hill	2.26	0.7	Man, Od, WB, Tri, Nag	Inceptisols, Entisols
Tarai	0.33	0.1	UP, Sikkim	Mollisols, Entisols
Brown forest	0.54	0.2	Kar, Mah	Mollisols, Inceptisols
Mountain meadow	0.06	–	J&K	Mollisols
Sub-montane	0.1	–	J&K	Alfisols
Others*	28.0	8.6	–	–
Total	328.7	100	–	–

Abbreviations for states and other regions: AP, Andhra Pradesh; A&N, Andaman and Nicobar; ArP, Arunachal Pradesh; Del, Delhi; Guj, Gujarat; Har, Haryana; J&K, Jammu and Kashmir; Kar, Karnataka; Ker, Kerala; MP, Madhya Pradesh; Mah, Maharashtra; Man, Manipur; Nag, Nagaland; Od, Odisha; Pb, Punjab; Pod, Pondicherry (now Puducherry); Raj, Rajasthan; TN, Tamil Nadu; Tri, Tripura; UP, Uttar Pradesh; WB, West Bengal.

Source: Bhattacharya *et al.* (2013)

5. 2:1 silicate minerals acquire negative charge due to isomorphic substitution of Si by Al or Mg and this charge is permanent in nature.
6. In tropical Ultisols and Oxisols, silicate minerals are not present in large amounts and the soil colloidal complex is dominated by hydroxides of Al and Fe. These Al and Fe hydroxides have colloidal properties and acquire negative charge due to OH^-, which depend on soil *p*H. This is why this negative charge is variable in nature. In acid conditions the charge is positive (+), while in alkaline conditions it is negative (–).
7. Soil organic matter (SOM) is an important constituent. It has colloidal properties and life elixir of soil biota. It is also store house of plant nutrients.
8. Soil is a living body and holds many macro (earthworms, insects), meso (mites, colembola etc.) and microfauna (bacteria, fungi, actenomycetes etc).
9. There are 11 major soil groups in India, namely, alluvial soils, coastal alluvial soils, red soils, black soils, laterite soils, desert soils, hill soils, *tarai* soils, brown forest soils, mountain meadow soils and sub-montane soils.

REFERENCES

Barton, C.D. and Karathanasis, A.D. 2002. Clay minerals. (In) *Encyclopedia of Soil Science.* Marcel Dekker, Inc., the USA.

Bhattacharya, T., Pal, D.K., Mandal, C., Chandran, P., Ray, S.K. , Sarkar, D., Velmourougane, K., Srivastava, A., Sidhu, G.S., Singh, R.S., Sahoo, A.K., Dutta, D., Nair, K.M., Srivastava, R., Tiwary, P., Nagar, A.P. and Nimkhedkar, S.S. 2013. Soils of India: historical perspective, classification and recent advances. *Current Science* **104**(10): 1308–1323.

Grewal, J.S. and Kanwar, J.S. 1976. Potassium and ammonium fixation in Indian soils – A review. Indian Council of Agricultural Research, New Delhi, 75 pp.

Keddy, P.A. 2010. *Wetland Ecology: Principles and Conservation*, edn 2, 497 pp. Cambridge University Press, Cambridge, the UK.

Krishnamoorhty, N.H., Goswami, C.L and Dayal, J. 1987. Effect of waterlogging and growth retardants on gram. *Indian Journal of Plant Physiology* **30**(4): 387–389.

Kumar, P., Pal, M., Joshi, R. and Siram, R.K. 2013. Yield, growth and physiological responses of mungbean [*Vigna radiata* (L.) Wilczek] genotypes to waterlogging at vegetative stage. *Physiology and Molecular Biology of Plants* **19**(2): 209–220.

Kumutha, D., Sairam, R.K., Ezhilmathi, K., Srivastava, G.C., Deshmukh, P.S. and Meena, R.C. 2009. Waterlogging induced oxidative stress and antioxidant activity in pigeonpea genotypes. *Biologia Plantarum* **53**(1): 75–84.

Lal, R. 2009. Soil degradation as a reason for inadequate human nutrition. *Food Science* **1**: 45–57.

Lal, R. 2014. Soil carbon sequestration for sustained nutrient supply. (In) *Textbook of Plant Nutrient Management,* pp. 48–63. Prasad, R., Kumar, D., Rana, D.S., Shivay, Y.S. and Tewatia R.K. (Eds.), Indian Society of Agronomy, New Delhi.

Misra, R.V., Roy, R.N. and Hiraoke, H. 2003. *On-farm Composting Methods.* Food and Agricultural Organization of the United Nations, Rome, 34 pp.

Oafoku, N.P., Ranst, E.V., Noble, A.D. and Baert, G. 2004. Variable charge soils: Their mineralogy, chemistry and management. *Advances in Agronomy* **84**: 159–215.
Prasad, R. and Power, J.F. 1997. *Soil Fertility Management for Sustainable Agriculture*, 356 pp. CRC Press, Boca Raton, FL, the USA.
Prasad, S., Ram, P.C. and Uma, S. 2004. Effect of waterlogging duration on chlorophyll content, nitrate reductase activity, soluble sugars and grain yield of maize. *Annals of Plant Physiology* **18**: 1–5.
Prasad, R., Shivay, Y.S. and Kumar, D. 2013. *A Practical Manual of Analytical Methods for Soil and Plant Samples from Agronomy Field Experiments*. Division of Agronomy, Indian Agricultural Research Institute, New Delhi, 58 pp.
Rao, C.N.R. 1999. *Understanding Chemistry*, 252 pp. University Press, Hyderabad.
Rao, D.L.N., Aparna, K. and Mohanty, S.R. 2019. Microbiology and biochemistry of soil organic matter, carbon sequestration and soil health. *Indian Journal of Fertilisers* **15**(2): 124–141.
Sanchez, P.A. 1976. *Properties and Management of Soils in the Tropics*, John Wiley, New York. 618 pp.
Saxena, K.G., Rao, K.S. and Maikhuri, R.K. 2014. Soil as a living body. (In) *Textbook of Plant Nutrient Management,* pp: 23–47. Prasad, R., Kumar, D., Rana, D.S., Shivay, Y.S. and Tewatia R.K. (Eds.), Indian Society of Agronomy, New Delhi.
Sharma, S.N., Prasad, R., Dwivedi, M.K., Kumar, S., Davari, M.R. and Shukla, L. 2010. Crop diversification and residue incorporation for making rice–wheat cropping systems sustainable. *Journal of Sustainable Agriculture* **34:** 342–364.
Shukla, M.K. 2013. *Soil Physics: An Introduction*. CRC Press, Boca Raton, FL, the USA.
Singh, S.B. and Abrol, I.P. 1988. Long-term effect of gypsum application and rice–wheat cropping on changes in soil properties and crop yield. *Journal of the Indian Society of Soil Science* **36:** 316–319.
Soil Survey Staff. 1999. *Soil Taxonomy: A basic system of soil classification for making and interpreting soil surveys*, Handbook 436, edn 2. Natural Resources Conservation Service, United States Department of Agriculture, Washington, DC.
Sorte, N.V., Rant, J.S., Doetale, R.D., Patel, M.C and Katekhaye, S.D. 1995. Effect of water logging on soybean of critical growth stages. *Journal of Soils and Crops* **5**(2): 141–144.
Subba Rao, N.S. 1999. *Soil Microbiology*, 407 pp. Enfield, N.H.-Science Publishers, Inc.

QUESTIONS

1. What is the difference between land and soil?
2. What is the difference between soil texture and soil structure?
3. What is the formula for determining pore space in soil using bulk density and particle density?
4. Name the five major soil groups of India.
5. What is variable charge?

Match the following

1. (i) Clay	(a) Particle size 0.002–0.02 mm
(ii) Coarse sand	(b) Particle size 0.02–0.2 mm
(iii) Silt	(c) Particle size 0.2–2 mm
(iv) Fine sand	(d) $\leq$ 0.002 mm.

2. (i)	Soil mesofauna	(a)	Earthworms
(ii)	Soil microfauna	(b)	Columbella
(iii)	Soil macrofauna	(c)	Nematodes

3. (1)	Vertisols	(a)	Soils with little development of soil profile
(2)	Mollisols	(b)	Soils with surface layer rich in Fe
(3)	Aridisols	(c)	Soils with surface layer rich in organic matter
(4)	Ultisols on drying	(d)	Soils rich in smectite minerals that crack
(5)	Entisols	(e)	Sandy desert soils

Answer True (T) or False (F)

(a) Largest area in India is under red soils.
(b) Cultivation is very difficult in black soils.
(c) Soil *p*H in laterites is generally below 6.0.
(d) Higher the content of kaolinite in a soil, higher is its CEC.
(e) Adding organic matter to a soil lowers its CEC.
(f) Isomorphous substitution takes place only in the tetrahedral layer of a 2 : 1 layer silicate
(g) Ionic radius of Si is larger than that of Al.
(h) The net charge on silicate minerals is affected by soil *p*H.
(i) The soils in the tropics have a positive charge when the *p*H is below 5.0.

Chapter 2

Historical Aspects of Soil Fertility

INTRODUCTION

In biology, fertility is defined as capacity to produce off-springs; however, in Soil Science a fine difference is made between a fertile and a productive soil. *Soil Productivity is the ability of a soil to produce a crop using the entire spectrum of its physical, chemical and biological attributes*, while *Soil Fertility refers to the capacity of a soil to supply all essential plant nutrients in the desired amounts*. Thus, soil fertility management is basically the nutrient management. To put it in simple words, a productive soil is always fertile, while a fertile soil may not be productive for a variety of factors, such as lack of adequate moisture in drier regions or lack of adequate air due to water-logging for most upland crops. Of late, a term '*Soil Health*' has been introduced. *Soil Health* is an assemblage of chemical, physical and biological parameters that closely relate to native or acquired production capacity and sustenance of ecologically important regulatory role (Katyal *et al.*, 2016). It is measured against predefined parameters of soil's status of fertility, compact ability, erodibility and biology. This again tends to relate soil fertility to nutrient content, availability and management.

HUMAN CREATIVITY, DEVELOPMENT OF AGRICULTURAL SYSTEMS AND SOIL FERTILITY

Although humans (*Homo sapiens*) evolved some 200,000 years ago, they remained a hunter-gatherer type until the agriculture (crop husbandry) developed some 13,000 to 10,000 years ago. However, before practicing agriculture, humans practiced *pastoralism*, where they herded the sheep and goats and moved from place to place in search of fertile soils having plenty of grass and other forages (Prasad *et al.*, 2016a). In some regions, the humans cleared forest and grew cereals and other crops for a period of 3–10 years and then abandoned the plant nutrient deprived land and moved to another clearing, which was fertile enough to grow crops. They would then return back to earlier clearing after its natural rejuvenation. This type of farming is known as shifting cultivation or *jhum*; it is still practiced in many parts of the world including north-east India (Shimrah *et al.*, 2015). This shows that both in pastoralism and shifting cultivation, humans were aware of the concept of a fertile soil. However, in addition to soil fertility, plenty of sunshine, water and warm weather were needed for growing good crops. These conditions were available in the river valleys of Tigris-Euphrates in Iraq (and adjoining countries), Indus valley in India, Yangtze-Yellow river valley in China and Nile valley in Egypt (Prasad *et al.*, 2016a). The river valleys of Tigris-Euphrates are known as Fertile Crescent and are considered the cradle of modern agriculture according to western literature (Bender, 1975). In these river valleys, animals (cattle, buffaloes, horses) were used for the cultivation and transportation of farm goods. The manure produced from animal dung and urine was applied to agricultural fields to maintain soil fertility. The cow urine contains 6.8–21.6 g N L^{-1}, of which 69% could be urea, while sheep and goat urine may contain 3.0–13.7 g N L^{-1} of which 83% could be urea (Bristow *et al.*, 1992). Thus, unknowingly some urea was applied to farm fields in Indus Valley period (3000–1300 BC), though urea was first manufactured in India only in 1959 at Sindri in Bihar (Tandon, 2014). In some parts of India, sheep and goat herders are paid to keep their herd on a farm field, because it improves soil fertility, which could partly be owing to urea added through sheep/ goat urine; the practice is known as penning.

Farmers in the Fertile Crescent found that growing beans (*Phaseolus* sp.), clover (*Medicago* sp.), and lupins (*Lupinus albus* L.) helped in restoring soil fertility and this led to the three-plot system of farming throughout Europe. In this system, the field was divided in 3 plots, one was sown with wheat (*Triticum aestivum*) or barley (*Hordeum vulgare*), the other with beans/ lupins/ clover, while the third was left as fallow. These plots were rotated in the subsequent years. The famous English Norfolk 4-year crop rotation practiced in the 18th century was: turnip (*Brassica rapa* subsp. *rapa*)–barley–clover–wheat (Martin *et al.*, 1988). Similarly in the USA, Thomas Jefferson followed a 5-year crop rotation of wheat/ potato (*Solanum tuberosum* L.)–pea (*Pisum sativum* L.)–rye (*Secale cereale* L.)/wheat–clover–buckwheat (*Fagopyrum esculentum* Moench) on his farm in Virginia (Karlen *et al.*, 1994). In India, the farmers in the rice–bowl (Eastern Uttar Pradesh, eastern Madhya Pradesh, Chhattisgarh, Bihar, Jharkhand and West Bengal) generally grew *khesari* (*Lathyrus sativus* L.) in the rice fallows (Prasad *et al*, 2016b). Archaeological studies have revealed the existence of *khesari* seeds in Bihar around 2000 BC (Mehra, 2000). However, it was only in 1886 of the Science era when Hellriegel and Wilfarth discovered the nitrogen-fixing capacity of legumes, and in 1901 Beijerinck isolated the N-fixing bacteria *Rhizobium* sp.

These facts show that farmers in the Fertile Crescent and elsewhere were concerned with the maintenance of soil fertility. Most of the agricultural systems and crop rotations just discussed were developed by humans owing to their intuition and creativity (Prasad *et al.*, 2016a).

DEVELOPMENT OF SCIENCE AND SEARCH FOR PRINCIPLE OF VEGETATION

Science based on measurements and establishing the relationship between 'cause and effect' started somewhere around 550 BC with Astronomy, when Copernicus and Galileo made predictions about planets; science matured by the end of 16 century (Bansal, 2014). It is after that; the efforts were made to study the 'principle of vegetation' in Europe. Some of the observations, made by the European scientists about the materials responsible for plant growth are given in Table 2.1.

Liebig in his report "Chemistry in its application to Agriculture and Physiology", submitted to the British Association in 1840, made the following important observations (Tandon, 2014).

1. Most of the C in plants comes from air, while water is taken from soil.
2. Phosphates are essential for seed formation.
3. Alkaline metals are needed for neutralization of acids formed by the plants.
4. Plants absorb everything from the soil indiscriminately but the roots release what is not needed by the plants.
5. Plants absorb N as ammonium, which they obtain from air, soil and manures.

Table 2.1: Materials responsible for plant growth as per observations

Material	Scientist
Water	British scientist Francis Bacon (1577–1624); Dutch scientist Jan Baptist van Helmont (1577–1644)
Salt peter (KNO_3)	German–Dutch scientist Johann Rudolf Glauber (1606–1668); British scientist John Mayow (1643–1679)
Soil particles	British scientist Jethro Tull, the discoverer of plough (1674–1741)
C from soil– Humus theory	British scientist Sir Humphry Davy (1778–1829); German Agronomist Albrecht Daniel Thaer (1752–1828)
C from air	French scientist John Baptiste Boussingault (1802–1888)
P from soil	German chemist Justus von Liebig (1803–1873)

Some of the above statements made by Liebig were later found to be incorrect, but he also enunciated the *Law of Minimum* in *1843, which states that an essential nutrient in the shortest supply in relation to plant needs, controls the yield of a crop*. Law of minimum holds true even today and has been the basis for the concepts of *Balanced Fertilization* and *Site-Specific Nutrient Management* (*SSNM*), which have played a role in fertilization of crops. It is now recognized that German botanist Carl Sprengel or Kurt Sprengel (1789–1859) had enunciated the *Law of Minimum* as early as 1828 (Tandon, 2014).

LONG-TERM FIELD EXPERIMENTS AND SOIL FERTILITY

Nicolas-Théodore de Saussure (1767–1845), a Swiss Chemist and Plant Physiologist, made observations that the plants absorbed N and a number of other minerals from soil set the stone rolling for the search of essential plant nutrients. His work was followed by the French Chemist Jean-Baptiste Joseph Dieudonné Boussingault (1801–1887), who established the first agricultural experiment station on his wife's property in Pechelbronn in Alsace, France - 60 km north of Strasbourg in 1836. His work was published in 8 volumes of *Agronomie, Chimie Agricole et Physiologie* (Boussingault, 1860–1891). His experimental station did not survive the vacillations of the 1870 Franco-Prussian war. Liebig acknowledged Boussingault as the pioneer and great discoverer of many advances in soil and plant chemistry. Rothamsted Experimental Station having been founded in 1843 at Harpenden, Hertfordshire, England is now the oldest agricultural research institutions in the world, while the Morrow Plots, named after Professor George E. Morrow, an experimental agricultural station at the University of Illinois at Urbana–Champaign, the USA, established in 1876, is the second oldest agricultural experimental station in the world.

SOIL FERTILITY RESEARCH IN INDIA

Famines have been a regular feature of India and millions of people died in these famines due to hunger and malnutrition because of lesser production and poor food management (Prasad, 2013). The then British Government constituted a Famine Commission in 1880, which recommended an improvement in Indian agriculture to overcome the food shortages. Poor soil fertility, especially the low organic matter content of Indian soils as compared to England, attracted the attention of British administrators and Dr John Walter Leather was appointed as the first Agricultural Chemist at the then Department of Agriculture at Dehradun in 1882. According to Ghosh (1984), Leather was the 'Father of Soil Science' in India. He carried out in-depth studies of soils of India and made extensive investigations on sodic soils of Uttar Pradesh.

In 1899–1900 famine, about 55 million people were affected and there were 1 million fatalities (Scott, 2017). To improve the agriculture in India, state departments of agriculture were established and later the Imperial (now Indian) Agricultural Research Institute (IARI) was established in 1905 at Pusa

(Bihar), which was later shifted, to New Delhi in 1935 due to the destruction of the institute building by an earthquake. Dr Leather was also transferred to the IARI at Pusa. The first soil map of India was made in 1943 by B. Vishwanath and A.C. Ukil at the IARI, New Delhi (Ghosh, 1984).

Even India's freedom in 1947 followed the aftermath of Bengal famine of 1943–44 (Goswami, 1990) and Dr A.B. Stewart from the Macaulay Institute of Soil Research of the University of Edinburgh was invited to suggest measures to improve soil fertility and food production in India. Dr Stewart suggested the conductance of trials on the farmer's fields (Simple Fertilizer Trials) as well as at research stations (named as Model Agronomic Experiments) for studying the response to chemical fertilizers (Stewart, 1947). Dr H.N. Mukherjee of the Department of Agriculture, Bihar, had already suggested conductance of trials on farmers' fields as a new method of determining nutritional requirements of crops (Mukherjee, 1952). The Indian Council of Agricultural Research, New Delhi, initiated a scheme of Simple Fertilizer Trials (SFT) in 1952–53. In addition, Model Agronomic Experiments Scheme was also initiated for conducting well-planned field experiments at selected research centres under the overall direction by the ICAR. Both these schemes were merged together to form the All India Coordinated Agronomic Research Project (AICARP) in 1968 (now named as AICRP on Integrated Farming Systems or AICRP-IFS) (PDFSR, 2012). At present, there are 25 main centres, 12-subcentres and 6 sub-subcentres under AICRP-IFS. In addition, there are 32 centres for conducting simple fertilizer trials. A few major findings from soil fertility and nutrient management viewpoint from this project and some other sources are as under:

1. Thousands of SFTs conducted all over the country before mid-1960s, due to tall Indian wheat and rice varieties grown then, showed that the response to N in most cases was limited to 30 kg N ha^{-1} and application of P and K did not yield good response (Pense and Khanna, 1964). The situation, however, changed after the introduction of high-yielding dwarf varieties of wheat (and rice) and good responses to N up to 90 kg N ha^{-1} or even more were obtained in trials under this project and at other research centres (Sharma and Singh, 1966; Mahapatra *et al.*, 1973; Prasad, 2009). Data of Sharma and Singh (1966) are given in Table 2.2. Numerous such data were generated all over the country, which changed the fertilizer scenario in India. A good response to P (Prasad *et al.*, 1971; Tandon, 1987; Prasad, 2007; NAAS, 2009) and K was also reported (Prasad and Mahapatra, 1970; Tiwari, 2014).
2. Cropping system-based production technologies were developed with special reference to, integrated nutrient management (INM), site-specific nutrient management (SSNM), precision agriculture, and water management.

3. Substitution of 50% of N by FYM / crop residue/ green manure to rice in the recommended NPK dose and application of recommended dose of NPK to wheat increased the soil organic carbon (SOC), microbial biomass carbon , and C–sequestration (for more, read Chapter 14 on organic manures).
4. Higher crop yields were recorded in long-term substitution of 25–50% N with FYM or green manure in all the major cereal–cereal cropping systems (CS); green manure was found to be better than FYM and crop residues in rice–rice CS, whereas in other cropping systems, FYM performed better than green manure.

Table 2.2: Grain yield (kg ha^{-1}) of dwarf Mexican and tall Indian wheat varieties as influenced by nitrogen fertilization

Wheat variety	0 kg N ha^{-1}	45 kg N ha^{-1}	90 kg N ha^{-1}	135 kg N ha^{-1}
Mexican dwarf varieties				
Sonora 63 (kg ha^{-1})	2,170	2,332	4,266	3,935
Sonora 64 (kg ha^{-1})	2,372	3,565	4,585	4,165
Lerma Rojo (kg ha^{-1})	1,735	3,889	4,774	4,085
Average (kg ha^{-1})	2,092	3,562	4,542	4,062
Agronomic Efficiency (AEn) (kg grain kg^{-1} N)	–	32.7	27.2	14
Tall Indian Varieties				
NP 876 (kg ha^{-1})	1,590	2,736	2,797	2,412
NP 887 (kg ha^{-1})	1,617	2,605	2,444	1,019
C 306 (kg ha^{-1})	2,034	2,730	2,765	2,032
Average (kg ha^{-1})	1,747	2,690	2,669	2,121
Agronomic Efficiency (AEn) (kg grain kg^{-1} N)	–	21	10.2	2.8

Source: Sharma and Singh (1966)

5. A most productive and profitable cropping systems atlas for different regions of the country has been prepared with appropriate nutrient-management package of practices.

Inspired by the Rothamsted Experimental Station, a number of long-term experiments were established in India at a number of locations, which included: Kanpur in Uttar Pradesh (1885); Pusa in Bihar (1908); Coimbatore in Tamil Nadu; Padegaon in Maharashtra; and Shahjahanpur in Uttar Pradesh. By 1980 all experiments had been abolished except the experiment at Coimbatore which is still continuing. However, a new series of Long-term Manurial Experiments have been started since 1971 at 17 centres under the aegis of the Indian Institute of Soil Science, Bhopal, Madhya Pradesh. Findings from these long-term experiments have helped in understanding several aspects of soil fertility and nutrient management including the changes in soil *p*H and the need for liming. Some of the main findings from these experiments are (Swarup *et al.*, 1998):

1. Continuous application of N fertilizer alone is the least desirable practice, as it leads to the depletion of other plant nutrients. At some centres continuously N fertilized plot yields were lower than the control plot yields. This proved the need for balanced fertilization and SSNM (site-specific nutrient management).
2. Soil health (good physical, chemical and biological properties) is maintained with annual addition of 10–15 tonnes of farmyard manure (FYM). This lays emphasis on integrated nutrient management (INM).
3. On acid soils liming helps in increasing crop yields.
4. There is need for monitoring available Zn and S and apply these where necessary.

SOIL TESTING

Soil testing as a tool for monitoring and managing soil-fertility status was initiated about 100 years back in USA (Tandon, 2014). In India, soil-testing scheme was launched in 1955–56 with 16 soil-testing laboratories. Today, there are about 1,275 state and central government-operated soil-testing laboratories in the country. In addition, several fertilizer-producing companies and as well private sectors run their own soil-testing laboratories, which are currently 226 in numbers. There is also an All-India Co-ordinated Research Project on Soil Test Crop Response (AICRP-STCR) with about 25 centres and an All-India Co-ordinated Research Project on Micro- and Secondary Nutrients and Pollutant Elements in Soils and Plants (AICRP-MSPE) with about 21 co-operating centres under the aegis of the IISS, Bhopal, to take care of micro-, secondary- and pollutant elements in soils. Thus, soil-fertility research in India is fairly well organized.

This is just a brief discussion on soil-fertility research in India. There are a large number of research projects underway on soil fertility under state governments, state agricultural universities and central agricultural institutes. Also, most fertilizer producers have a fairly large number of demonstrations promoting nutrient use in their marketing areas.

SUMMARY

1. India has a history of famines and adequate food production has always been a priority for Government of India since the times of British Raj.
2. Soil-fertility research in India is now fairly well organized and a good progress has been made, which has enabled efficient fertilizer

management for most cropping system of the country. It has helped in increasing fertilizer consumption from a mere 0.55 tonnes in 1950–51 to 25.95 million tonnes (MMT) in 2016–17 (16.74 MMT N, 6.71 MMT P_2O_5 and 2.50 MMT K_2O).

REFERENCES

Bansal, B. 2014. *The Birth of Science & Evolution of Science.* Published by the Author, the USA.

Bender, B. 1975. *Farming in Pre-History-From Hunter Gatherer to Food Production*, 268 pp. Billings & Sons, London.

Boussingault, Jean-Baptiste (1860–1891). *Agronomie, Chimie Agricole et Physiologie* (in French), edn 2. 8 Volumes, Mallet-Bachelier, then Gauthiers-Villars, Paris.

Bristow, A.W., Whitehead, D.C. and Cockburn, J.E. 1992. Nitrogenous constituents in the urine of cattle, sheep and goats. *Journal of the Science of Food and Agriculture* **59**(3): 387–394.

Ghosh, A.B. 1984. History of development of agriculture in India with special reference to Soil Science during 1800–1950. (In) *Soil Science in India, Bulletin of Indian Society of Soil Science* **14**: 14–28.

Goswami, O. 1990. The Bengal famine of 1943: Re-examining the data. the *Indian Economic and Social History Review* **27**: 445–463.

Karlen, D.L., Varvel, G.E., Bullock, D.G. and Cruse, R.M. 1994. Crop rotations for the 21st century. *Advances in Agronomy* **53**: 1–45.

Katyal, J.C., Dutta, S.P. and Golui, D. 2016. Global review on state of soil health. (In) *Soil Health–Concept, Status and Monitoring,* Katyal J.C., Chaudhari S.K., Dwivedi B.S., Rattan R.K., and Majumdar K (Eds.). *Bulletin of the Indian Society of Soil Science* **30**: 1–33.

Mahapatra, I.C., Prasad, R., Krishnan, K.S., Goswami, N.N. and Bapat, S.R. 1973. Response of rice, *jowar*, maize, *bajra*, groundnut and castor to fertilizers under rainfed conditions on farmers' fields. *Fertiliser News* **18**(8): 18–28.

Martin, J.H., Leonard, W.M. and Stamp, D.L. 1988. *Principles of Field Crop Production*, edn 3, Macmillan, New York.

Mehra, K.L. 2000. History of crop cultivation in pre-historic India. (In) *Ancient and Medieval History of Indian Agriculture and its Relevance to Sustainable Agriculture in the 21 Century.* Choudhary S.L., Sharma G.S. and Nene Y.L. (Eds.), Rajasthan College of Agriculture, Udaipur, pp. 1–21.

Mukherjee, H.N. 1952. Soils of Bihar and a new method of determining their manurial requirements. *Proceedings Bihar Academy of Agricultural Science,* India **1**: 58–94.

NAAS. 2009. Crop Response and Nutrient Ratio. National Academy of Agricultural Sciences, New Delhi. Policy Paper No. 42.

PDFSR. 2012. *PDFSR at a Glance.* Project Directorate for Farming Systems Research, Modipuram, Meerut, Uttar Pradesh. https://www.researchgate.net/publication/284142718_PDFSR_at_a_Glance.

Pense, V.G. and Khanna, R.C. 1964. Response of some crops in India to fertilizers and factors influencing this response. *Indian Journal of Agricultural Sciences* **34**(3): 172–204.

Prasad, R. 2007. Phosphorus management in the rice–wheat cropping system of the Indo-Gangetic plains. *Better Crops India* **1**: 8–11.

Prasad, R. 2009. Efficient fertilizer use: The key to food security and better environment. *Journal of Tropical Agriculture* **47**(1–2): 1–17.

Prasad, R. 2013. Population growth, food shortages and ways to alleviate hunger. *Current Science* **105**: 32–36.

Prasad, R. and Mahapatra, I.C. 1970. Crop responses to potassium on different soils. *Fertiliser News* **15**(2): 48–56.

Prasad, R. Bhendia, M.L. and Turkhede, B.B. 1971. Relative efficiency of phosphatic fertilizers in different soils of India. *Proceedings of the International Symposium* on *Soil Fertility Evaluation*, New Delhi, **1**: 747–756.

Prasad, R., Nene, Y.L., Singh, G. and Shivay, Y.S. 2016b. Pulses from ancient through present day India. *Indian Journal of Agronomy* **61**(4th IAC Special issue): S19–S32.

Prasad, R., Shivay, Y.S. and Nene, Y.L. 2016a. Asia's contribution to the evaluation of agriculture: creativity, history and mythology. *Asian Agri-History* **20**(4): 233–250.

Scott, J.E. 2017. *In Famine Land: Observations and Experiences in India, During the Great Drought of 1899–1900* (Classic Reprint-Paperback), Forgotten Books, London, UK, 298 pp.

Sharma, K.C. and Singh, V. 1966. Response of Mexican wheat to nitrogen. *Fertiliser News* **11**(9): 18–22.

Shimrah, T., Rao, K.S. and Saxena, K.G. 2015. The shifting agricultural systems (*Jhum*) and strategies for sustainable agroecosystems in North-east India. *Agroecology and Sustainable Food Systems* **39**: 1,154–1,171.

Stewart, A.B. 1947. *Soil Fertility Investigations in India.* Indian Council of Agricultural Research, New Delhi.

Swarup, A., Reddy, D.D. and Prasad, R.N. (Eds.). 1998. *Long-term Soil Fertility Management Through Integrated Plant Nutrient Supply,* 335 pp. Indian Institute of Soil Science, Bhopal, Madhya Pradesh.

Tandon, H.L.S. (Ed.). 1987. *Phosphorus Research and Agricultural Production in India,* 160 pp. Fertilizer Development and Consultation Organization, New Delhi.

Tandon, H.L.S. 2014. History of soil fertility. (In) *Textbook of Plant Nutrient Management,* pp. 1–22. Prasad, R., Kumar, D., Rana, D.S., Shivay, Y.S. and Tewatia, R.K. (Eds). Indian Society of Agronomy, New Delhi.

Tiwari, K.N. 2014. Potassium management. (In) *Textbook of Plant Nutrient Management,* pp. 122–143. Prasad, R., Kumar, D., Rana, D.S., Shivay, Y.S. and Tewatia, R.K. (Eds), Indian Society of Agronomy, New Delhi.

QUESTIONS

1. Which British scientist advised the Government of India to conduct simple fertilizer trials on the farmers' fields? He came from which Institute in UK?
2. In which year soil testing was started in India.
3. What are four major findings from the Long–term Fertilizer Experiments?

Match the followings

(1)	Carbon from soil	(a)	Glauber
(2)	Water is the principle of vegetation	(b)	Francis Bacon
(3)	Salt peter is the principle of vegetation	(c)	Justus V. Liebig
(4)	Carbon from air	(d)	Bossingault
(5)	Phosphorus is taken from soil	(e)	Humphrey Davy

Answer True (T) or False (F)

(a) Soil fertility and soil productivity are synonyms.
(b) Law of minimum was propounded by Sir Humphrey Davy.
(c) First agricultural research station was started at Rothamsted in UK.
(d) J.W. Leather was the first Agricultural Chemist appointed in India.
(e) J.W. Leather started his work at Pusa, Bihar.

Chapter 3

Essential and Beneficial Plant Nutrients

German chemist Justus Freiherr von Liebig (1803–1873), Professor at the University of Geissen and the French chemist Jean-Baptiste Joseph Dieudonne' Boussingault (1801–1887), Professor at the University of Lyon and later Chair, Conservatoires des Artes et Matier in Paris are to be credited for analyzing the plants and coming with the idea that plants absorbed nitrogen and other mineral nutrients from the soil for their growth. D.I. Arnon and P.R. Stout, Plant Physiologists and Plant nutritionists at the University of California, Berkeley, USA, grew plants by hydroponics using pure chemical compounds and established following 3 criteria for the essentiality of nutrients for plant growth (Arnon and Stout, 1939).

1. Plants cannot complete their life-cycle when a particular element is deficient in soil.
2. The deficiency symptoms are specific to that element.
3. The element is essentially involved in the metabolism of the plant (This is not applicable to beneficial elements).

So far 18 elements have been declared essential for plant growth. These are C, H, O, N, P, K, Ca, Mg, S, Fe, Mn, B, Zn, Cu, Mo, Cl, Ni and V. Of these 18 elements, C is taken from CO_2 in the air and H and O from water in the soil. Fifteen essential plant nutrients that are to be monitored in soil and managed properly for good crop production are listed in Table 3.1. These elements are, in general, classified as primary, secondary and micronutrients.

Table 3.1: Essential plant nutrients that are taken from soil or foliage (when sprayed)

Class Element	Primary form in which taken up	Year of Discovery	Discovered by
Primary			
Nitrogen (N)	NH_4^+, NO_3^-	1804	De Saussure
		1851–55	Boussingault
Phosphorus (P)	$H_2PO_4^-$, $HPO_4^=$	1839	Liebig
		1861	Ville
Potassium (K)	K^+	1866	Birner and Lucanus (1866)
Secondary			
Calcium (Ca)	Ca^{++}	1862	Stohmann

Class Element	Primary form in which taken up	Year of Discovery	Discovered by
Magnesium (Mg)	Mg^{++}	1865	Sachs (1965)
		1875	Boehm
Sulphur (S)	$SO_4^{=}$	1865	Sachs (1865)
Micronutrients			
Iron (Fe)	Fe^{++}	1843	Gris (1843)
Manganese (Mn)	Mn^{++}	1922	McHargue (1922)
Copper (Cu)	Cu^{++}	1925	McHargue (1925)
Boron (B)	H_3BO_3	1923	Warrington
Zinc (Zn)	Zn^{++}	1926	Somner and Lipman (1926)
Molybdenum (Mo)	$MoO_4^{=}$	1939	Somner and Lipman (1926)
Chlorine (Cl)	Cl^-	1954	Arnon and Stout (1939)
Nickel (Ni)	Ni^{++}	1987	Broyer *et al.* (1954)
Vanadium (V)	VO_3-	1973	Brown *et al.* (1987) Welch and Huffman (1973)

Source: Barker and Pilbeam (2007); Traceable references are provided.

CLASSIFICATION OF ESSENTIAL PLANT NUTRIENTS

On the basis of amounts taken up by crop plants

Essential plant nutrients are classified on the basis of amounts taken up by crop plants. There are 2 major classes, namely (i) macronutrients and (ii) micronutrients. However, for practical purposes, the macronutrients are further classed into primary and secondary nutrients.

Primary plant nutrients: These nutrients include N, P and K and are taken up by crop plants in the largest amounts (Table 3.2). The uptake of these nutrients is generally expressed in kg tonne^{-1} (tonne or 1,000 kg). These nutrients are to be applied each crop season and form the basis of fertilizer industry in the world.

Secondary plant nutrients: These nutrients include Ca, Mg and S and are also removed in fairly large amounts, but are not the primary products of the fertilizer industry. The uptake of these nutrients is also generally expressed in kg tonne^{-1} (tonne or 1,000 kg). Sulphur deficiency is wide spread in India and S fertilization of some crops, such as, rapeseed-mustard and pulses has become necessary in most regions of the country.

Micronutrients: These nutrients include Zn, Fe, Mn, Cu, B, Mo, Cl, Ni (nickel) and V (vanadium). The uptake of these nutrients is generally expressed in g or mg tonne^{-1} (metric tonne or 1,000 kg) (Table 3.3). Zinc deficiency is wide spread in India and Zn fertilization of most crops is needed in most regions of the country. Nutrients Ni and V have been recently included and are needed in very small amounts in some special crops.

Table 3.2: Total nutrient uptake (kg t^{-1}) of macronutrients (primary and secondary) by some crops

Crop group	Crop	N	P	K	S	Ca	Mg
Cereals	Rice	20.0	4.8	25.0	3.0	7.0	3.0
	Wheat	25.0	3.9	27.5	4.7	5.3	4.7
	Maize	29.9	5.9	27.3	–	–	–
Pulses	Chickpea	60.7	4.0	32.7	8.7	18.7	7.3
	Pigeonpea	70.8	6.6	13.3	7.5	19.2	12.5
Oilseeds	Rapeseed	32.8	7.1	34.8	17.3	42.0	8.7
	Soybean	70.7	13.4	48.0	6.7	14.0	7.6
	Groundnut	58.1	8.5	25.1	7.9	20.5	13.3

Source: Prasad (2007).

All uptake data in this text are in tonne (metric ton or Mg or 1,000 kg; abbreviated as 't').

Table 3.3: Total uptake (g t^{-1}) of micronutrients by some crops

Crop Group	Crop	Zn	Fe	Mn	Cu	B	Mo
Cereals	Rice	40	153	675	18	15	2
	Wheat	56	624	70	24	48	2
	Maize	130	1,200	320	130	–	–
Pulses	Chickpea	38	868	70	11	–	–
	Pigeonpea	32	1,200	107	26	–	–
Oilseeds	Rapeseed	100	1,123	95	17	–	–
	Soybean	77	346	83	30	–	–
	Groundnut	109	2,284	93	36	–	–

Source: Prasad (2007)

On the basis of mobility in plants

Classification on the basis of mobility of nutrients in plants is important, because it indicates the plant parts, where one has to look for the deficiency symptoms. There are 2 classes:

Mobile nutrients: These include N, P, K, Mg, Mo and Cl. Deficiency symptoms of these nutrients appear on older/lower leaves, because in the case of deficiency nutrients from older leaves move to younger leaves.

Less mobile nutrients: These include Ca, S, Fe, Zn, Mn, Cu and B. Deficiency symptoms of these nutrients appear on younger leaves or on growing tips or fruits, because there is very little translocation from older leaves.

On the basis of biochemical and physiological roles in plants

Based on their biochemical and physiological functions, Mengel and Kirkby (1987) divided essential plant nutrients in 4 groups as given below:

A. Major constituent of organic plant components (carbohydrates, proteins, fats): C, H, O, N and S.
B. Nutrients involved in biochemical reactions, such as, esterification: P and B.
C. Elements present in free ionic state or are adsorbed to in diffusible organic anions (such as, carboxyl): K, Ca, Mg, Mn and Cl.
D. Elements predominantly present as chelates: Fe, Cu, Zn and Mo.

The division between group C and D is not clear cut, since Mg, Mn and Ca may also be present as chelated compounds.

FUNCTIONS OF ESSENTIAL PLANT NUTRIENTS

Different essential plant nutrients have specific functions in plants (Prasad *et al.*, 2016). These are briefly described.

Nitrogen

Nitrogen (N) is the number one plant nutrient. It is the nutrient applied in the largest amount as fertilizer. It has the following functions in plants:

1. As a constituent of chlorophyll it has a key role in photosynthesis, which is the primary process in the production of food.
2. It is a key element of proteins that separates them from carbohydrates and fats. Nitrogen is responsible for the production of plant proteins, which leads to the development and growth of shoot. Also N is responsible for the accumulation of protein in grains.
3. As a constituent of enzymes, N controls all physiological processes in plants.
4. As a constituent of nucleo-proteins (RNA, DNA), N plays a key role in heredity.

Phosphorus

1. As a constituent of adenosine triphosphate (ATP) phosphorus (P) plays a key role in energy transfer in photosynthesis.
2. As a constituent of nucleotide adenosine diphosphate (NADP), P is involved in the electron transfer in photosystem-I of photosynthesis.
3. It acts as a substrate for the synthesis of RNA and is an integral part of RNA and DNA.
4. This nutrient is a constituent of phospholipids in plants.

5. It is reported to promote root growth.
6. Phosphorus hastens maturity in crops.

Potassium

1. Potassium (K) has a key role in maintaining cell turgidity.
2. It is responsible for opening the stomata for photosynthesis.
3. It plays a role in the synthesis of ATP.
4. Potassium has a role in the transmission of photosynthates from source (leaves) to sink (organs).
5. It imparts resistance to a number of plant diseases.

Calcium

1. Calcium (Ca) is responsible for the rigidity of cell-walls.
2. It plays an important role in cell division and cell elongation.
3. This plant nutrient is important for the membrane stability and reduces its permeability.
4. It imparts selectivity to ion absorption and transport.
5. Calcium regulates soil *p*H, which controls nutrient availability to plants.

Magnesium

1. Magnesium (Mg) is the central cation holding 4 porphyrin rings in the structure of chlorophyll and has thus an important role in photosynthesis.
2. It has a role in sucrose transport in phloem.
3. Magnesium has a role in N metabolism.

Sulphur

1. Sulphur (S) as a constituent of amino acids viz. methionine and cysteine has a major role in protein synthesis.
2. It is a component of thiamine (Vitamin B-1) and biotin (Vitamin-H).
3. Sulphur is a component of a number of important S-compounds in plants, including, Alliin (which produces the pungent allicin, when garlics and onions are chopped), glucosinolates (the pungent secondary metabolites in the plants of *Brassicae* family), and glutathione (an important anti-oxidant in plants) (Prasad, 2014).

Zinc

1. As a component of enzyme carbonic anhydrase (CA) Zinc (Zn) is involved in carbohydrate metabolism. In the case of severe Zn deficiency, excretion of sugars at the leaf surface can be observed in some plants.
2. As a component of enzyme alcohol dehydrogenase (ADH), it plays an important role in anaerobic root respiration in rice.

3. As a component of enzyme Cu–Zn superoxide dismutase (Cu–Zn–SOD) it helps in detoxification of superoxide radicals.
4. Zinc is involved in the synthesis of RNA.
5. Zinc fingers play a role in gene expression.
6. In addition to above, Zn activates a large number of enzymes.

Iron

1. Iron (Fe) is a component of haem (Fe-containing) enzymes, catalases, peroxidases and cytochrome oxidases.
2. As a component of catalases, Fe plays a role in detoxification of free O_2 radicals.
3. This plant nutrient is involved in the formation of a-amino-laevulinic acid (ALA) a precursor of porphyrins.
4. As a component of peroxidases, it is involved in the transient change of Fe^{2+} to Fe^{3+} in the porphyrin rings.
5. As a component of ferrodoxins (having Fe–S clusters) it acts as in the redox system in photosynthesis.
6. Iron has a role in protein metabolism, nucleic acid metabolism and lipid metabolism.

Manganese

1. Manganese (Mn) is involved in the Hill reaction (splitting of water molecule into O_2 and H_2) in photosystem-II of photosynthesis.
2. It is responsible for the generation of O_2 in the Earth's atmosphere.
3. Manganese superoxidase (Mn-SOD) is responsible for protection of cells in the mitochondria from free O radicals.
4. Manganese plays a role in the synthesis of lignin, which imparts disease resistance in plants.

Copper

1. Copper (Cu) is responsible for electron transfer in photosystem-I of photosynthesis.
2. As a component of superoxide dismutase (Cu–Zn–SOD), it is responsible for protection of chloroplast from free O radicals.
3. It plays a role in the synthesis of lignin, which imparts disease resistance in plants.

Boron

1. Boron (B) has a role in the structure of cell-walls.
2. Boron has a role in pollen germination and pollen-tube growth and therefore in determining the quality of flowers and fruits.
3. It has a role in the synthesis of RNA.

Molybdenum

1. Molybdenum (Mo) is involved in nitrogen fixation in legumes.
2. It is involved in nitrate reduction.
3. Molybdenum is involved in the catabolism of S-containing amino acids in the mitochondria.

Chlorine

1. Chlorine (Cl) is important in water relations in plants.
2. It has a major role in stomatal regulation. In oil palm trees it can lead to wilting.
3. Chlorine is the dominant anion in the vacuole.

Nickel

1. Nickel (Ni) has a role in ureolysis, hydrogen metabolism and methane biogenesis.

Vanadium

1. Vanadium (V) has a role in the synthesis of d-aminolevulinic acid (d-ALA) in the autotrophically growing green alga *Chlorella pyrenoidosa*.

DEFICIENCY SYMPTOMS OF PLANT NUTRIENTS

Deficiency symptoms differ from crop to crop. Some general guidelines are given below:

Mobile nutrients (*symptoms appear on older leaves*)

Entire or nearly entire leaf affected

Phosphorus deficiency: Stunted growth, older leaves tending to be dark green, purple or reddish complete leaf or margins.

Nitrogen deficiency: Stunted growth, older leaves pale green to yellowish and dying.

Molybdenum deficiency: Light green to yellow necrotic spots; leaves may be scorched, cupped or rolled. In young cauliflower plants, the deficiency symptoms are chlorosis of leaf margins and the whole leaves may turn white. The leaf blades do not develop properly. When the deficiency is severe, only the midribs develop, which is called as *Whiptail disease*.

Only part of the leaf affected

Mg deficiency: Interveinal chlorosis, pale/ yellow stripes, red or dead spots.

K deficiency: No interveinal chlorosis, burned leaf margins and tips.

Cl deficiency: No interveinal chlorosis, chlorotic/ necrotic lesions, wilting of leaves, especially at margins is a typical symptom of Cl deficiency. However, Cl toxicity is more common a problem than Cl deficiency. Most fruit trees, berries and vines are sensitive to high concentration of Cl (> 5 g kg^{-1} dry matter). Premature yellowing and abscission of leaves are other symptoms of Cl toxicity.

Less mobile nutrients (*symptoms appear on young leaves and growing points*)

Growing points affected

Boron deficiency: Boron (B) deficiency symptoms include dark green, leathery, downward cupping of young leaves and die-back of shoot tips. The known B deficiencies include hollow heart in groundnut (*Arachis hypogaea* L.) and formation of 'cork like' material in the curd of cauliflower (*Brassica oleracea* var. *botrytis* L.). Boron is known as a fruit quality nutrient.

Calcium deficiency: Brown, twisted, dying growing point, young leaves may stick to the immediate lower leaf. Some of the known Ca deficiencies are bitter pit in apple (*Malus pumila* L.), blossom end rot in tomato (*Solanum lycopersicum* L.) and black head in celery (*Apium graveolens* L.).

Younger leaves affected

Sulphur deficiency: Pale green/ yellow young leaves, stunted weak plants. In rapeseed-mustard and other plants of the *Brassicae* family, cupping of the leaves may take place. In chickpea (*Cicer arietinum* L.), Sulphur (S) deficiency develops in the middle trifoliates.

Zinc deficiency: Pale green/yellowish middle leaves in maize (*Zea mays* L.); in extreme deficiency in maize the growing point may have yellowish white young leaves (known as white bud). In citrus (*Citrus* sp.) young leaves may form a *rosette* (a bunch of small leaves).

In rice and wheat brown spots appear on younger and middle leaves; the brown spots may coalesce to cover the entire leaf (*khaira disease*). Burned brown spots appear in the field.

Iron deficiency: Interveinal chlorosis giving a fine reticulate pattern in leaves; in extreme deficiency the young leaves may turn white as in sorghum [

Sorghum bicolor (L.) Moench]. A spray of 0.5% ferrous sulphate solution will turn these leaves green in a couple of days. This is real test for iron deficiency. In waterlogged lowland paddy Fe and Mn toxicity can occur, which results in developing of yellow-brown spots on leaves.

Manganese deficiency: Interveinal chlorosis and yellow spots on leaves in dicots, while in monocots, such as oats (*Avena sativa* L.), grey specks appear in leaves. In oats, Mn deficiency produces a condition called '*grey speck*' which occurs in patches; plant turns pale green and young leaves have spots or lesions of grey/ brown necrotic tissue with orange margins.

Copper deficiency: Bushy plants with twisted dried tip; impaired ear/ panicle formation.

BENEFICIAL PLANT NUTRIENTS

Beneficial plant nutrients that are not essential for all plants, but have been reported to improve the growth of some plants and in some cases have been reported to enhance their resistance to biotic and abiotic stresses (Prasad *et al.*, 2016). Presently five beneficial plant nutrients are recognized. These include: sodium (Na), silicon (Si), cobalt (Co), aluminium (Al) and selenium (Se) (Simranjeet-Kaur *et al.*, 2016). A brief discussion on these follows.

Sodium

1. Sodium (Na) can partly substitute for K in plants for two mechanisms: *(i)* osmotic and electrical balance in cells and *(ii)* regulation of functioning of stomata.
2. Abundant supply of sodium is required for good growth of sugar-producing crops, such as fodder and sugarbeet (*Beta vulgaris* L.).
3. In C_4 plants, Na plays a role in the ultra structure of chloroplasts and in the conversion of pyruvate to phosphoenol pyruvate.
4. In addition to its role in plants, adequate Na concentration is required in herbage to overcome Na deficiency in grazing cattle (Na is an essential nutrient for men and animals).

Silicon

1. Adequate supply of Silicon (Si) has been reported necessary for the normal growth of rice, sugarcane, barley and beet (Prasad and Power, 1997).
2. Good yield responses to an application of sodium/ calcium silicate have been obtained in rice and sugarcane (Savant *et al.*, 1997).
3. Silicon is essential for a group of algae known as diatoms and for the plants of the *Equisetaceae* family (scouring rushes).

4. Imparts resistance to blast disease and insect attack in rice.
5. Helps plants to overcome the salinity and drought stresses.
6. Silicon helps plants to overcome the toxicity to Mn and Al.
7. Addition of Si increases the availability of P in some soils.
8. Silicon helps plants to overcome damage due to ultraviolet exposure.

Cobalt

1. Cobalt (Co) is involved in N-fixation by legumes and a good response of subterranean clover (*Trifolium subterraneum* L.) to Co fertilization was reported from Australia (Ozanne *et al.*,1963).
2. Responses of a number of crops including cowpea [*Vigna unguiculata* (L.) Walp.], Faba bean (*Vicia faba* L.), groundnut, tomato to Co fertilization have been reported.
3. Cobalt as a constituent of Vitamin B12 (Cobalamin) is essential for humans and animals and they cannot synthesize it. Even plants cannot synthesize Cobalamin. Only bacteria an *archaea* can synthesize it (Prasad *et al.*, 2016).

Selenium

1. Certain algae require Selenium (Se) to produce selenoproteins.
2. Some species of *Astralagus*, *Stanleya*, *Morinda*, *Neptunia*, *Oonopsis* and *Xylorhiza* are known as Se accumulators (Terry *et al.*, 2000).
3. In recent years, Se has received attention in human nutrition and is included in mineral supplements.

Aluminium

1. Not much information on Aluminium (Al) nutrition of plants is available.
2. Some plant species accumulate Al, which are mainly found in tropical regions and with few exceptions are woody plants. Physiological mechanisms underlying the uptake of Al into above-ground plant tissues have been investigated in buckwheat or kutu (*Fagopyrum esculentum* L.), Indian rhododendron (*Melastoma malabathricum* L.), big leaf hydrangea [*Hydrangea macrophylla Hand-Mazz.*] and tea [*Camellia sinensis* (L.) Kuntze], but the results are not clear (Schmitt *et al.*, 2016).
3. The Al-induced increase in the activities of antioxidant enzymes, resulting in increased membrane integrity and delayed lignifications and aging, can be considered as a possible reason for the stimulatory effects of Al on the growth of the tea plants (Ghanati *et al.*, 2005).

SUMMARY

1. There are 18 essential plant nutrients, namely C, H, O, N, P, K, Ca, Mg, S, Fe, Mn, Zn, Cu, B, Mo, Cl, Ni and V. Out of these, plants take C from CO_2 in the atmosphere and H and O from soil water. The rest 15 elements have to be applied to soil in addition to the native amounts that soils can supply as per crop needs. The materials supplying these are known as fertilizers. Soil-fertility management deals with the availability of these 15 elements.
2. The 15 essential elements needed to be applied to soils as per crop needs are classified into primary, secondary and micronutrients.
3. Primary plant nutrients are 3 and include N, P and K and are taken up by crop plants in the largest amounts and have to be applied each crop season. The uptake of these nutrients is generally expressed in kg $tonne^{-1}$ (metric ton or 1,000 kg). These nutrients form the basis of fertilizer industry in the world.
4. Secondary plants nutrients are also three and include Ca, Mg and S and are also removed in fairly large amounts, but are not the primary products of the fertilizer industry. The uptake of these nutrients is also generally expressed in kg $tonne^{-1}$. Calcium needs of crops on acid soils are met by liming. Sulphur deficiency is wide spread in India and S fertilization of some crops, such as, rapeseed-mustard and pulses has become necessary in most regions of the country.
5. Micronutrients are nine and include Zn, Fe, Mn, Cu, B, Mo, Cl, Ni and V. The uptake of these nutrients is generally expressed in g or mg $tonne^{-1}$.
6. Essential plant nutrients are also classified on the basis of their mobility in plants as mobile and less mobile.
7. Mobile nutrients include N, P, K, Mg, Mo and Cl. Deficiency symptoms of these nutrients appear on older/ lower leaves, because in the case of deficiency nutrients from older leaves move to younger leaves.
8. Less-mobile nutrients include Ca, S, Fe, Zn, Mn, Cu and B. Deficiency symptoms of these nutrients appear on younger leaves or on growing tips or fruits.
9. In addition to essential plant nutrients, there are five beneficial plant nutrients, namely Na, Si, Co, Se and Al, which are required by some crop plants. Sodium (Na) can partly substitute for K. Silicon has been reported beneficial for rice and sugarcane. Cobalt is involved in nitrogen fixation in legumes. The Selenium (Se) is a component of selenoproteins. The Al has some stimulatory effect on the growth of tea plants.
10. Deficiency of an element leads to specific deficiency symptoms in plants, which can be overcome by in-time soil or foliar application of that nutrient.

REFERENCES

Arnon, D.I. and Stout, P.R. 1939. Molybdenum as an essential element for higher plants. *Plant Physiology* **14**(3): 590–592.

Arnon, D.I. and Stout, P.R. 1939. The essentiality of certain elements in minute quantity for plants with special to copper. *Plant Physiology* **14**: 371–375.

Barker, A.Y. and Pilbeam, D.J. 2007. *Handbook of Plant Nutrition*. CRC Press, Boca Raton, FL, the USA.

Birner, H. and Lucanus, B. 1866. Wassrculturversuche mit hafer. *Landw Vers Sta* **8**: 128–177 (in German).

Brown, P.H., Welch, R.M. and Cary, E.F. 1987. Nickel: A micronutrient essential for higher plants. *Plant Physiology* **85**(3): 801–803.

Broyer, T.C., Carlton, A.B., Johnson, C.M. and Stout, P.R. 1954. Chlorine–a micronutrient element for higher plants. *Plant Physiology* **29**(8): 526–532.

Ghanati, F., Morita, A. and Yokota, H. 2005. Effects of aluminium on the growth of tea plant and activation of antioxidant system. *Plant and Soil* **276**(1–2): 133–141.

Gris, E. 1843. Memoir relatif a laction de composes soluble ferruguineaux sur la vegetation. *Compte Rendu de l'Academie de Sciences* **17**: 679 (in French).

McHargue, J.S. 1922. The role of manganese in plants. *Journal of American Chemical Society* **44**(7):1,592– 1,598.

McHargue, J.S. 1925. The significance of occurrence of copper, manganese and zinc in forage crops and foods. *Journal of American Agronomy Society* **17**: 368.

Mengel, K. and Kirkby, E.A. 1987. *Principles of Plant Nutrition*, 686 pp. Panima Book Agency, New Delhi.

Ozanne, P.G., Greenwood, E.A.N. and Shaw, T.C. 1963. The cobalt requirement of subterranean clover in field. *Australian Journal of Agricultural Research* **14**: 39–50.

Prasad, R. 2007. *Crop Nutrition–Principles and Practices*, 272 pp. New Vishal Publications, New Delhi.

Prasad, R. 2014. Major sulphur compounds in plants and their role in human nutrition and health–An overview. *Proceedings of Indian National Science Academy* **80**(5): 1045–1054.

Prasad, R. and Power, J.F. 1997. *Soil Fertility Management for Sustainable Agriculture*, 356 pp. CRC Press, Boca Raton, FL, the USA.

Prasad, R., Majumdar, K., Shivay, Y.S. and Kapil, U. 2016. *Minerals in Plant and Human Nutrition and Health*, 80 pp. International Plant Nutrition Institute (IPNI), Peachtree Corners, GA, USA.

Sachs, J. 1865. *Handbuch der Experimental–Pysiologie der Pflanzen*, 528 pp. Wilhelm Engelmann, Leipzig, Germany. (in German).

Savant, N.K., Synder, G.H. and Datnoff, L.E. 1997. Silicon management and sustainable rice production. *Advances in Agronomy* **58**: 151–199.

Schmitt, M., Watanabe, T. and Jansen, S. 2016. The effects of aluminum on plant growth in a temperate and deciduous aluminum accumulating species. *AoB Plants* **8:** plw065. (doi: 10.1093/aobpla/plw065).

Simranjeet-Kaur, Navneet-Kaur, Siddique, H.M. and Nayyar, H. 2016. Beneficial elements for agricultural crops and their functional relevance in defense against stresses. *Archives of Agronomy and Soil Science* **62**(7): 905–920.

Somner, A.L. and Lipman, C.B. 1926. Evidence of the indispensable nature of zinc and boron for higher plants. *Plant Physiology* **1**(3): 231–249.

Terry, N., Zayed, A.M., De Souza, M.P. and Tarun, A.S. 2000. Selenium in higher plants. *Annual Review of Plant Physiology and Plant Molecular Biology* **51**: 401–432.

Welch, R.M. and Huffman (Jr), E.W.D. 1973. Vanadium and plant nutrition–The growth of lettuce (*Latuca sativa* L.) and tomato (*Lycopersicum esculentum* Mill.) plants in nutrient solution low in vanadium. *Plant Physiology* **52**: 183–185.

QUESTIONS

1. What are three criteria for the essentiality of a plant nutrient?
2. What are the number of micronutrients essential for plants?
3. According to Mengel and Kirkby, which 5 essential plant nutrients are major constituent of organic plant components.
4. Match the nutrients in Series A with a function in Series B.

SeriesA: (i) S, (ii) P, (iii) Fe, (iv) Zn, (v) N, (vi) K, (vii) B, (viii) Mn, (ix) Mg, (x) Ca, (xi) Cu, (xii) Mo, (xiii) Cl

Series B: *(a)* component of proteins, *(b)* methionine, *(c)* energy transfer, *(d)* component of chlorophyll, *(e)* cell turgidity, *(f)* cell integrity, *(g)* plays an important role in anaerobic root respiration in rice, *(h)* is involved in the formation of a-amino–laevulinic acid (ALA), *(i)* is responsible for the generation of O_2 in the Earth's atmosphere, *(j)* has a role in sucrose transport in phloem, *(k)* electron transfer in photosystem I of photosynthesis, *(l)* role in the quality of fruits *(m)* is involved in nitrogen fixation in legumes, *(n)* a major role in stomatal regulation.

Match the followings

(a)	Zinc	(i) *Whiptail* disease
(b)	Mn	(ii) Cork-like material in the curd of cauliflower
(c)	B	(iii) Purple coloration of leaves
(d)	Mo	(iv) *Khaira* disease of rice
(e)	P	(v) Grey speck disease of oats

Answer True (T) or False (F)

(a) Zinc, iron, Mo and N are mobile nutrients.
(b) K and Mo are less mobile nutrients.
(c) S deficiency appears on lower leaves.
(d) B is known as a fruit quality nutrient.
(e) Balanced fertilization in India refers to a balanced application of N, P and Zn fertilizers.

Fig. 3.1: Deficiency symptoms of some nutrients in plants.
(a) N deficiency in maize (causing dying lower leaves); (b) P deficiency in sugarcane (resulting purpling of leaves, poor tillering); (c) K deficiency in wheat (dying top leaves); (d) Zn deficiency in wheat resulting interveinal chlorosis in young (*left*) leaves, yellow-brown spots (*right*); (e) Yellowing of young leaves with dead margins due to S deficiency in sugarcane; (f) B deficiency in pineapple (Fruit necrosis which begins with the browning of inner most part of the mesocarp).
(*Source:* Vikaspedia–A Government of India portal)

Chapter 4

Carbon, Soil Organic Matter and Carbon Sequestration

All life on planet Earth is carbon centered; hereditary characters are controlled by DNA and RNA, growth by proteins and activities by energy from carbohydrates and fats – all these are carbon compounds. Not only this, but the life also ends as carbon compounds CO_2 or CH_4, depending on aerobic or anaerobic decomposition. Incidentally soil fertility also depends very much on soil organic carbon (SOC). In ancient days, farmers used to apply manure to maintain soil fertility and even now, urea, the nitrogen fertilizer used in the largest amounts, is a carbon compound. All living forms on the Earth including all plant and animal species and their activities are a part of global carbon cycle. The **carbon cycle** is the biogeochemical cycle by which carbon is exchanged among the biosphere, pedosphere, geosphere, hydrosphere, and atmosphere of the earth.

According to a University of New Hampshire study (UNH, 2008), the amount of C in different pools is as follows: *1*, Earth's core ~100 million Pg C (1 Pg=10^{15} g =10^9 metric tonnes or 1 billion tonnes or Mg) mostly as sedimentary rocks; *2*, Fossil fuels ~4,000 Pg C; *3*, Oceans ~38,000 Pg C; *4*, Surface soil ~1,500 Pg as SOC; additional 950 Pg C as inorganic C mostly carbonates (Yousaf *et al.*, 2016); *5*, Plants including forests ~560 Pg C; *6*, Earth's atmosphere ~750 Pg C, mostly as CO_2, but some as CH_4 and other C compounds. Bhattacharya *et al.* (2000) reported that SOC (soil organic C) stock in 1.5 m soil depth in India was 29.92 Pg C, while SIC (soil inorganic C) stock was 33.98 Pg C. Its break up in soils of different soil orders in India is shown in Table 4.1. Srinivasarao *et al.* (2014) reported that in semi-arid regions of India, the magnitude of SOC, SIC and total C was 2.9, 1.9 and 4.8 Pg C, respectively.

As regards, SOC stock in forest soils in India, it is estimated at 4.13 Pg C in the surface 50 cm depth and at 6.81 Pg C in the surface 1 m depth (Chhabra *et al.*, 2003). For the Shiwalik hill region of India, Saha *et al.* (2014) reported the following SOC stock values in the 1.05 m layer surface soils under different ecosystems: 83.5 Mg ha^{-1} for forest soils, 73.5 Mg ha^{-1} for grasslands, 61.1 Mg ha^{-1} for cultivated soils and 55.6 Mg ha^{-1} for eroded soils. Thus, cultivated soils have lesser SOC than that of grasslands and forests.

Table 4.1: Stock of soil organic C (SOC) and soil inorganic C (SIC) in different soil orders in India

Soil order	SOC (Pg C)	SIC (Pg C)	Total (Pg C)
Entisol	2.56	2.86	5.42
Vertisol	8.77	6.14	14.90
Inceptisol	5.81	7.04	12.85
Aridisol	2.02	1.40	15.42
Mollisol	0.49	0.07	0.56
Alfisol	9.72	4.48	14.20
Ultisol	0.55	0.00	0.55
Total	29.92	33.98	63.90

Source: Bhattacharya *et al.* (2000)

PHYSICAL CHARACTERIZATION OF SOIL ORGANIC MATTER

The soil organic matter (SOM) has been classified in four classes, which are as under:

Crop residues

Particles of greater than 2 mm.

Particulate organic matter

Particles less than 2 mm and greater than 0.053 mm in size. Particulate Organic Matter (POM) is biologically and chemically active and is part of the labile (easily decomposable) pool of soil organic matter (SOM) (Cambardella and Elliot, 1992). The POM is generally higher in no-till cultivation and increases with the addition of organic manures (Fronning *et al.*, 2008).

Humus

Particles of < 0.053 mm. A major part of the advantages of SOM in improving soil fertility, holding and releasing of nutrients and storing of soil moisture are attributed to this fraction of SOM (Berg and McClaugherty, 2007).

Recalcitrant or resistant organic C (ROC)

The recalcitrant or resistant organic carbon (ROC) are biologically stable; typically as charcoal.

LABILE FRACTIONS OF SOIL ORGANIC CARBON

Particulate organic matter (POM) and humus form the labile fractions of SOC. These fractions decompose fast and are responsible for soil fertility and the release of plant nutrients. A number of procedures have been proposed for determining different fractions of labile SOC. These include POM

(Cambardella and Elliot, 1992), hot water-soluble carbon (HWSC) (Schulz *et al.*, 2003), 333 m M $KMnO_4$-extractable C (KMOC) (Blair *et al.*, 1995), Walkley and Black C (WBC) (Walkley and Black,1934) and microbial biomass C (MBC) (Jenkinson and Powlson, 1976). Verma *et al.* (2013) reported that, in intensively cultivated alluvial soils of north India, WBC, KMOC and MBC constituted on average 25, 10 and 2% of the SOC pool. Saha *et al.* (2014) reported that, in surface 0–15 cm soil POM constituted 75% of SOC in forest soils, 65.1% of SOC in eroded soils, 64.7% in grassland soils and 55.3% in cultivated soils in the Shiwalik hill region of India. As regards KMOC, the values were 15.7% for forest soils, 15.4% for forest soils, 10.7% for cultivates soils and 8.2% for eroded soils. Similarly, the values for HWSC were 8.2% for eroded soil, 6.45% for forest soils, 6.4% for cultivated soils and 5.8% for grassland soil. These data show that SOC in different labile fractions do not have the same components. In general, forest and grassland soils had higher labile C than the cultivated soils.

CARBON-SEQUESTRATION

Climate model projections summarized in the IPCC 2013 report indicated that, during the 21st century, the global surface temperature is likely to rise a further 0.3 to 1.7°C (0.5 to 3.1°F) in the lowest emissions scenario, and 2.6 to 4.8°C (4.7 to 8.6°F) in the highest emissions scenario (IPCC, 2013). This could lead to floods, cyclones and rise in sea-levels (Ramsay and Sobel, 2011) and great sea-shore cities, such as, Bombay (*now* Mumbai), Calcutta (*now* Kolkata), and Chennai, New York, London etc. are at the greatest risk. On the other hand, global warming can lead to droughts in some areas (Trenbath *et al.*, 2014). These changes are caused by global warming gases, carbon dioxide (CO_2), methane (CH_4) and nitrous oxide (N_2O), of which CO_2 is the dominant one. Atmospheric CO_2 concentration has increased from 280 ppmv (parts per million on volume basis) in the pre-industrial era of 1750s to 400 ppmv in 2015; however, a major increase was observed in 20th century, which is the highest for the past 800,000 years (UESPA, 2017). Global warming is increasing at fairly fast rate in the 21th Century since all but one of the 16 hottest years in NASA's 134-year record have occurred since 2000 (MacMillan, 2016). Major cause for the increase in atmospheric CO_2 is considered to be anthropogenic in nature, including deforestation, use of coal in generating electricity and other industries and burning of fossil fuel in vehicle etc.

Green plants can play a major role in absorbing the excess CO_2. About 123 Pg C year^{-1} is taken out of the atmosphere by terrestrial plant photosynthesis (Gross Primary Productivity or GPP) out of which 60 Pg C year^{-1} is returned back to the atmosphere by plant respiration, leaving a balance of 63 Pg C year^{-1} added to Earth globally (Net Primary Productivity or NPP). However, an equivalent 53 Pg C year^{-1} is respired from soil. This gives a net ecosystem productivity (NEP) of 10 Pg C year^{-1}. Out of this 5–7 Pg C year^{-1} may be lost by fire and perturbations, leaving about 3–5 Pg C year^{-1} as net biomass productivity (NBT) (Lal, 2014).

All C sequestered by plants finally finds its way to soil as organic matter, either directly as root exudates and root residues, leaf fall and incorporation of crop residues after the harvesting of grains or other economic products or indirectly as manure from the animals feeding on the plant foliage or as solid waste or urine from humans feeding on plant and animal food. This is why C-sequestration by plants is important from the soil health viewpoint in addition to its role in reducing global warming.

AGRONOMIC PRACTICES AND C SEQUESTRATION IN INDIA

Pathak *et al.* (2011) reported that in 17 out of 26 sites under the Long-term Fertilizer Experiments, an application of FYM (~10 t ha^{-1} year^{-1} during *kharif* season) along with recommended dose of NPK increased organic C in soil by 0.65% over the control as compared to an increase of only 0.16% in NPK plots. The average carbon-sequestration potential (CSP) over 17 years of study in different sites ranged from 2.1 to 4.8 Mg C ha^{-1} [Mg is same as tonne (t) or 1,000 kg]. Conservation agricultural practices can certainly affect the amount of C sequestration. In a recent study at New Delhi, Parihar *et al.* (2018) reported that after a 5 year of maize-based cropping systems, total SOC in conventional tillage plots was only 0.88 Mg C ha^{-1} as compared to 7.22 Mg C ha^{-1} in zero-tilled plots. In *tarai* region of Nepal, following the rice–wheat system with chemical fertilizer (NPK) alone for a period of 25 years reduced the native organic matter pool of forest soil by 47%, however, application of farmyard manure @ 4 tonnes per year with NPK, helped in recovering the SOC loss by 71–76% (Gami *et al.*, 2009). Some data on the effects of cropping systems and manurial practices in India on C-sequestration are shown in Table 4.2.

Table 4.2: Long-term effects of different nutrient management on C sequestration rate (0–15 cm soil depth) under different cropping systems

Cropping system management	Nutrient	C sequestration (kg ha^{-1})	Soil type
Rice–wheat–jute (30 years)	NPK	18.2	Typic Entrochrept
	NPK + FYM	50.7	
Sorghum–wheat (15 years)	NPK	189.0	Typic Haplustert
	NPK + FYM	295.0	
Maize–wheat–cowpea (31 years)	NPK	84.0	Typic Haplustept
	NPK + FYM	149.0	
Maize–wheat–cowpea (34 years)	NPK	380.0	Typic Haplustept
	NPK + FYM	513.0	
Soybean–wheat–maize (38 years)	NPK	146.0	Typic Haplustert
	NPK + FYM	333.0	
Rice–fallow (10 years)	FYM	110.0	Aeric Endoaquept
	FYM + GM	90.0	
	FYM + GM + crop residue	120.0	

Source: Manna *et al.* (2019)

CROPPING AND AGRO-FORESTRY SYSTEMS AND CARBON SEQUESTRATION

Srinivasarao *et al.* (2014) reported that, under semi-arid conditions long-term cropping without adding organic matter can severely deplete SOC, and depletion was the most under groundnut– finger millet [*Eleusine coracana* (L.) Gaertn.] system (0.92 Mg C ha^{-1} $year^{-1}$). Agroforestry systems have received special attention in this connection. Sharma *et al.* (2015) reported that in the rice–wheat cropping system in north India, when poplar (*Populus* sp.) was planted on boundaries of fields, C sequestration was 1.03 Mg C ha^{-1} $year^{-1}$ and it increased to 3.18 Mg C ha^{-1} $year^{-1}$, when rice straw was incorporated in soil and it further increased to 4.64 Mg C ha^{-1} $year^{-1}$, when both rice and wheat straw were incorporated in soil. Srinivasarao *et al.* (2014) observed that, agro-forestry systems involving *Eucalyptus* and *Leucaena* plantations can sequester to the extent of 10 Mg C ha^{-1} $year^{-1}$. There is projected 630 million ha potential area for agro-forestry, which could sequester 0.586 million Mg C $year^{-1}$ by 2040 (Nair, 2012). Potential of soil C sequestration in India is estimated at 39 to 49 (44 ± 5) Tg C $year^{-1}$ (Lal, 2004).

EXPRESSIONS USED FOR CALCULATING SOME IMPORTANT TERMS IN RELATION TO SOC

Soil C budget (Lal, 2014)

ΔSOC input ($C_r + R_b + M_c + D_p$) – C losses (D + E + P)

where Cr is crop residue, R_b is root biomass, M_c is manure or compost, D_p is deposition, D is decomposition, E is erosion and P is percolation loss of dissolved organic matter carbon (DOC). When C input > C losses, *SOC is sequestrated* and if C losses > C input, *SOC is depleted.*

Carbon management index (CMI) (Zhao *et al.*, 2014)

From the soil fertility point of view, labile C is important and crop-management practices that increase the content of labile C improve soil fertility. A carbon-management index (CMI) has been suggested to measure the C maintenance efficiency of a crop-management system.

Firstly a C pool index (CPI) is calculated, then a lability index (LI) and finally the CMI is calculated as per expressions below.

CPI= Sample SOC (g kg^{-1})/ Reference sample SOC (g kg^{-1})

[Reference sample could be the soil sample at the beginning of a cropping system or a sample from the soil which is taken as the standard]

LI= L (Lability) in sampled soil/ L (Lability) in the reference soil,

where, L= Content of labile C / Content of non-labile C.

Any lability index (hot-water soluble C, 333 m M $KMnO_4$–extractable C or microbial biomass) can be used. Finally CMI is calculated as:

CMI= CPI × LI × 100

SUMMARY

1. Earth's core contains ~100 million Pg C.
2. Globally surface 1 m soil ~1,500 Pg C as SOC, while the atmosphere contains ~750 Pg C.
3. Indian soils contain ~30 Pg C as SOC and ~34 Pg C as SIC in surface 1.5 m layer.
4. Forest soils have more SOC than cultivated soils.
5. The SOC has four classes: crop residue-particles > 2 mm; particulate organic matter (POM) > 0.053 < 2.0 mm; humus < 0.053 mm; resistant organic C (ROC).

6. The POM and humus from the labile fractions of SOC and is the most important from the soil fertility point of view.
7. Different fractions of labile SOC include: POM, hot water-soluble carbon (HWSC), 333 m M $KMnO_4$-extractable C (KMOC), Walkley and Black C (WBC) and microbial biomass C (MBC).
8. Green plants can play a major role in absorbing the excess CO_2. About 123 Pg C year^{-1} is taken out of the atmosphere by terrestrial plant photosynthesis (Gross Primary Productivity or GPP) out of which 60 Pg C year^{-1} is returned back to the atmosphere by plant respiration, leaving a balance of 63 Pg C year^{-1} added to earth globally (Net Primary Productivity or NPP).
9. Addition of FYM can increase C sequestration by soil.
10. Conservation agriculture also helps in C-sequestration.

REFERENCES

Berg, B. and McClaugherty, C. 2007. *Plant litter: Decomposition, Humus Formation, Carbon Sequestration,* edn 2, 338 p. Springer.

Bhattacharya, T., Ray, S.K., Pal, D.K., Velayuthaum, M., Chandran, P., Mondal, C. and Wani, S.P. 2000. Total carbon stocks in Indian soils: issues, priorities and management. (In) *International Seminar on Land Resource Management for Food, Employment and Environmental Security (ICLRM)*, 8–13 November 2000, New Delhi, Soil Conservation Society of India, pp. 1–46.

Blair, G.J., Lefroy, R.D.B. and Lisle, L. 1995. Soil carbon fraction based on their degree of oxidation and the development of a carbon management index for agricultural systems. *Australian Journal of Soil Research* **46**: 1,459–1,466.

Cambardella, C.A. and Elliot, E.T. 1992. Particulate soil organic matter changes across a grassland cultivation sequence. *Soil Science Society of America Journal* **56**: 777–783.

Chhabra, A., Palria, S. and Dudhwal, V.K. 2003. Soil organic carbon pool in Indian forests. *Forest Ecology Management* **173**(1–3): 187–199.

Fronning, B.E., Thelen, K.D. and Min, D. 2008. Use of manure, compost, and cover crops to supplant crop residue carbon in corn stover removed cropping systems. *Agronomy Journal* **100**: 1703–1710.

Gami, S.K., Lauren, J.G. and Duxbury, J.M. 2009. Soil organic carbon and nitrogen in Nepal long term fertilizer experiments. *Soil and Tillage Research* **106**(1): 95–103.

IPCC. 2013. *Climate Change 2013: The Physical Science Basis-Technical Summary*. PDF. pp. 89–90.

Jenkinson, D.S. and Powlson, D.S. 1976. The effect of biocidal treatment of metabolism in soil. V. A method for measuring soil biomass. *Soil Biology and Biochemistry* **8**: 209–213.

Lal, R. 2014. Soil carbon sequestration for sustained nutrient supply. (In) *Textbook of Plant Nutrient Management,* pp. 48–63. Prasad, R., Kumar, D., Rana, D.S., Shivay, Y.S. and Tewatia, R.K., (Eds), Indian Society of Agronomy, New Delhi.

Lal, R. 2004. Soil carbon sequestration in India. *Climatic Change* **65**(3): 277–296.

Manna, M.C., Wanjari, R.H., Singh, M., Patra, A.K. and Chaudhari, S.K. 2019. Long-term effect of balanced fertilizer, manure and amendment on yield sustainability and soil health. *Indian Journal of Fertilisers* **15**(2): 142–148.

McMillan, A. 2016. Global Warming 101. Natural Resources Defense Council (Nrdc.org), March 11, 2016.

Nair, P.K.R. 2012. Carbon sequestration studies in agroforestry systems: a reality check. *Agroforestry Systems* **86**: 243–253.

Parihar, C.M., Parihar, M.D., Sapkota, T.B., Nanwal, R.K., Singh A.K., Jat S.L., Nayak, H.S., Mahala, D.M., Singh, L.K., Kakraliya, S.K., Stirling, C.M. and Jat, M.L. 2018. Long-term impact of conservation agriculture and diversified maize rotations on carbon pools and stocks, mineral nitrogen fractions and nitrous oxide fluxes in Inceptisol of India. *Science of the Total Environment* **641**: 1,382–1,392.

Pathak, H., Byjesh, K., Chakrabarti, B. and Aggarwal, P.K. 2011. Potential and cost of carbon sequestration in Indian agriculture: estimates from long–term field experiments. *Field Crops Research* **120**(1): 102–111.

Ramsay, H.A. and Sobel, A.H. 2011. Effects of relative and absolute surface temperature on tropical cyclone potential intensity using a single-column model. *Journal of Climate* **24**: 183–193.

Saha, D., Kukal, S.S. and Bawa, S.S. 2014. Soil organic carbon stock and fractions in relation to land use and soil depth in the grassland Shiwalik hills of lower Himalayas. *Land Degradation and Development* **25**:407–416.

Sharma, R., Chauhan, S.K. and Tripathi, A.M. 2015. Carbon sequestration potential in agroforestry system in India: an analysis for carbon project. *Agroforestry Systems* (DOI: 10.1007/s10457–015-9840-8).

Srinivasarao, Ch., Lal, R., Kundu, S., Babu, M.B., Venkateswarlu, B. and Singh, A.K. 2014. Soil carbon sequestration in rainfed production systems in the semi-arid tropics in India. *Science of the Total Environment* **487**: 587–603.

Trenbath, K.E., Dai, A., Shrier, G., Jones, P.D., Barichirich, J., Biffa, K.R. and Sheffield, J. 2014. Global warming and changes in drought. *Nature Climate Change* **4**: 17–22.

UNH. 2008. Global Carbon Cycle– Pools, fluxes and a word about units, University of New Hampshire, Durham, NH 03824, USA.

USEPA. 2017. Climate Change Science–Causes of Climate Change-Search Snapshot 1/19/17(via internet).

Verma, B.C., Datta, S.P., Rattan, R.K. and Singh, A.K. 2013. Labile and stabilized fractions of soil organic carbon in some intensively cultivated soils. *Journal of Environmental Biology* **34**: 1,069–1,075.

Walkley, A. and Black, C.A. 1934. An examination of Degtjareff methods for determining soil organic matter and a proposed modification of the chromic acid titration method. *Soil Science* **37**: 29–38.

Yousaf, B., Liu, G., Wang, R., Abbas, Q., Imtiaz, M. and Liu, R. 2016. Investigating the biochar effects on C- mineralization and sequestration of carbon in soil compared with conventional amendments using stable isotope (d13C) approach. *Global Change Biology Bioenergy* (doi:10.1111/gcbb.12401).

Zhao, F., Yang, G., Han, X., Feng, Y., and Ren, G .2014. Stratification of carbon fractions and carbon management index in deep soil affected by the grain-to-green program in China. *PLoS ONE* **9**(6): e99657. https:// doi.org/10.1371/journal.pone.0099657.

QUESTIONS

1. Fill in the blanks

(a) The carbon cycle is the biogeochemical cycle by which carbon is exchanged among the biosphere, ..., geosphere, ..., and atmosphere of the Earth.

(b) Green plants can sequester about .... Pg C $year^{-1}$ from the atmosphere by photosynthesis.
(c) Global surface temperature is likely to rise a further .... °C in the lowest emissions scenario as predicted by IPCC.
(d) Total SOC in earth's surface is .... Pg C.
(e) One Pg is equal to .... tonnes (t).

2. Define the terms GPP, NPP, CPI, DOC, POC and LI

3. Answer True (T) or False (F)

(a) Labile C is only a fraction of total SOC.
(b) From the viewpoint of Soil Fertility, a soil having less labile soil C and high total SOC is better than a soil with high labile soil C and less total SOC.
(c) Contribution of root exudates to SOC is negligible and can be ignored

Chapter 5

Soil Hydrogen, Oxygen and Water

SOIL WATER

Soil water and oxygen share the pore space in soil and generally the oxygen content is determined by the water content, which is added by irrigation or rains and then keeps on changing due to evapotranspiration. Water is the vehicle by which most nutrients move from the soil to root surface. But for the lowland rice, nutrients are most available to plants at field capacity moisture (about 0.3 atmosphere soil water tension) and the availability decreases as the moisture content decreases to wilting point (about 15 atmosphere tension). That is why rainwater or irrigation increases nutrient availability, while extremely dry condition decreases it. There is thus a significant interaction between water and nutrient availability in soil (Singh *et al.*, 2014). In a study in India (Singh *et al.*, 1980) the yield of wheat was significantly affected by irrigation water and nitrogen treatments; maximum yield was obtained with irrigation at 50% available soil water in 60 cm soil depth and 120 kg ha^{-1} nitrogen addition. A study on wheat at Delhi (Kumar *et al.*, 2015) very well shows the effect of irrigation water on yield and uptake of nutrients (NPK) (Table 5.1).

Rice is special; it grows well in standing water. Flooding has also significant effect on soil *p*H, which itself has a significant influence on the availability of plant nutrients. Water management is crucial for overcoming Fe toxicity in rice (Sahrawat, 1979; Devi *et al.*, 1996).

SOIL AIR

But for lowland rice, oxygen is necessary for root respiration, which is also essential for nutrient uptake, especially of Fe, Zn and P. Linn and Doran (1984) showed that respiration and nitrification was at peak at 50% water filled pore space (WFPS), while denitrification started at 60% WFPS and was at peak at 100% WFPS. From crop growth and soil fertility point of view good aeration is required for:

1. Root growth and respiration–important for absorption of nutrients.
2. Microorganism population and activity affecting aerobic decomposition of SOM for the release of nutrients.

3. Necessary for nitrification.
4. Maize and most pulses are susceptible to prolonged waterlogging (*Refer* Chapter 1).
5. Preventing the formation of toxic gas H_2S and global warming gas CH_4.
6. For overcoming the S, Fe and Mn toxicity in lowland rice soils. This is achieved by creating periodic drainage.
7. Prevention of plant diseases, such as, wilt of gram and dieback of citrus (*Citrus* sp.) and peach [*Prunus persica* (L.) Botsch].
8. Oxidation of S applied as elemental S or pyrite as the source of S.
   a. In a 96-day incubation study at 25°C at 70% field capacity moisture with 10 soils, Zhao *et al.* (2014) found that S oxidation rate varied from 5.1 to 51.7 µg S cm^2 day^{-1}.

Table 1: Effect of irrigation on the grain yield and nitrogen, phosphorus and potassium uptake by wheat

Irrigation schedule	No. of irrigations	Grain yield (t ha^{-1})	N uptake (kg ha^{-1})	P uptake (kg ha^{-1})	K uptake (kg ha^{-1})
CRI	1	3.33	69.4	19.8	100.6
CRI + flowering	2	3.67	81.4	26.8	116.3
CRI + flowering + tillering + jointing + flowering + grain filling	6	4.31	102.9	35.8	170.7
LSD (P=0.05)		0.14	3.2	1.3	14.0

Source: Kumar *et al.* (2015); CRI–crown root initiation

SOIL HYDROGEN

Along with C and O, hydrogen is the chief element of all organic tissue of the living beings on Earth. From the soil fertility viewpoint, H^+ concentration in soil water is the deciding factor for the availability of plant nutrients in a soil. The H^+ concentration in soil solution is expressed as *p*H. The concept of *p*H was first introduced by the Danish chemist Søren Sørensen at the Carlsberg Laboratory in 1909. The *p*H is defined as the decimal (to base 10) logarithm of the reciprocal of the H^+ activity (aH^+) in a solution (Covington, 1985).

$$pH = -\log_{10}(aH^+)$$

The self-ionization of water is an ionization reaction in pure water or in an aqueous solution, in which a water molecule H_2O deprotonates (loses one of its H atom) to become a hydroxide ion OH^-. The H^+, so released immediately protonates another water molecule to form hydronium, H_3O^+. This can be expressed as under:

$H_2O + H_2O \rightarrow H_3O^+ + OH^-$

The ionization constant (Kw) of water can be expressed as:

$K_W = [H_3O]^+ [OH]^-$; the value of this constant is 14.

At neutral *p*H, the concentrations of both $[H^+]$ and $[OH^-]$ are 10^{-7}.

The *p*H scale is from 0 to 14. Values below 7 indicate acidity, while values above 7 indicate alkalinity. The ideal *p*H for the availability of most nutrients is 6.5. The availability of plant nutrients depends very much on soil *p*H. In general, there are four groups of soils.

Acid soils having *p*H below 6.5

Availability of micronutrients Fe, Mn, Zn and Cu increases as the *p*H goes down (soils become more acidic).

Soils near neutrality having *p*H 6.5–7.5

All primary and secondary nutrients are well available (N, P, K, Ca, Mg, S). Other nutrients are also available.

Alkaline soils having *p*H 7.5–8.5

Availability of Ca and B is high.

Strongly alkaline soils having *p*H above 8.5

Availability of micronutrients Mo and B increases with *p*H.

Soil *p*H is mostly determined in a soil : water suspension of 1 : 2.5. However, *p*H value will change on increasing the soil : solution ratio. For this reason some researchers prefer to determine soil *p*H in an electrolyte solution, such as, those of 1 M KCl or 0.01 M $CaCl_2$ (Prasad *et al*, 2006).

Soil *p*H indicates the active $[H^+]$ (active acidity) and is an intensity measure, just like temperature for heat.

The amount of lime ($CaCO_3$) required for neutralizing active acidity can be calculated as:

Neutralization reaction $CaCO_3 + 2HCl = CaCl_2 + H_2O + CO_2$

As per the above reaction each g (or kg) H will require 50 g (or kg) $CaCO_3$

The amount of $CaCO_3$ required to neutralize active acidity in a soil of *p*H 5 or concentration of $[H^+]$ of 10^{-5} M and having 20% moisture can be calculated as:

kg $CaCO_3 = [2.2 \times 10^6] \times [20 \times 10^{-2}] \times [10^{-5}] = 220$ kg

where $[2.2 \times 10^6]$ kg is the weight of 0–15 cm layer of one ha soil, $[20 \times 10^{-2}]$ refers to 20% soil moisture and $[10^{-5}]$ is the H^+ concentration in moles per liter (or kg) soil solution/ water.

This is a very small amount. However, when this amount of $CaCO_3$ is applied and the active acidity is neutralized, more $[H^+]$ will be released from the soil-exchange complex. This tendency of soil to resist a change in its *p*H is known as *buffering capacity*. Therefore for determining soil's *exchangeable-H capacity* (*known as lime requirement*), a buffer solution is used. Exchangeable-H is a *capacity factor* as compared to soil *p*H, which is an *intensity factor*. Exchangeable-H is also referred to as *potential acidity* as compared to the term *active acidity* for soil *p*H. A number of buffer solutions have been suggested by different researchers (Prasad and Power, 1997), for determining exchangeable-H, but the SMP (Shoemaker-McLean-Pratt) buffer made up of p-nitrophenol, triethanolamine, potassium chromate and calcium chloride adjusted to *p*H 7.5 is being used in India (Prasad *et al.*, 2006). Incidentally in India, rice is the principal crop grown on acid red and lateritic soils and rice soils are generally flooded. Flooding itself increases soil solution *p*H (Ponnamperuma, 1972) and therefore response to lime application in acid soils in India is limited, but it is obtained in Fe-toxic laterite soils of Asom, Jharkhand and Odisha (Jena and Rout, 2014). Interestingly, flooding lowers *p*H in sodic soils and rice is recommended as the recovery crop (Singh and Abrol, 1988). Lime is certainly needed for the pulses and other crops grown on rice fallows. Upland crops respond well to liming on acid soils. For example, the data in Table 5.2 from acid soils of Sylhet region in Bangladesh show a good response of wheat to liming.

Table 5.2: Effect of liming on grain yield t ha^{-1} of different wheat varieties on acidic soil of Sylhet, Bangladesh (2010–11)

Varieties	Lime	No lime
Shatabdi	3.98	3.32
Bijoy	4.01	3.60
Sourav	4.14	3.63
Sufi	3.82	3.13
Prodip	4.39	3.64
BARI Gom–25	4.06	3.46
BARI Gom–26	4.42	4.06
Mean	3.98	3.32
LSD (P=0.05)	0.41	

Source: Ataur Rahman *et al.* (2013); Soil *p*H was 4.8 and liming materials used was calcium oxide.

A number of materials are used for liming and their liming potential is generally given in terms of calcium carbonate equivalent (CCE of calcium carbonate is taken as 100). The CCE of different materials used for liming is as follows: calcium oxide 179; calcium hydroxide 136; dolomite 109; and basic slag 86.

As early as 1947, Mukherjee *et al.* (1947) had suggested that exchangeable Al was the major ion in the 1 N $BaCl_2$ extracts of acid soils and was the main cause of soil acidity. McLean *et al.* (1965) observed that, various forms of soil acidity appear to have the following order of activity: exchangeable or permanent charge H^+ > exchangeable or permanent charge Al^{3+} > hydroxyl–Al monomers > hydroxyl Al polymers = organic matter acidity = lattice Al–OH or Si–OH acidity. Since Al toxicity is the major cause of poor crop growth (Foy, 1992) in acid soils, some researchers (Sanchez, 1976) have suggested exchangeable Al as a criterion for determining lime requirement and it is followed in Brazil.

Soil *p*H also controls the activity of soil microorganisms. In a study having a *p*H range of 8.3 to 4.0, Rousk *et al.* (2009) found that bacteria prefer above neutral soil conditions, while the activity of fungi increases with an increase in acidity and most microbial activity tends to stop below *p*H 4.5. Nitrification is rapid under neutral to alkaline conditions (Sharawat, 2008); however, at *p*H higher than 8.0 oxidation of nitrite may be reduced resulting in nitrite accumulation (Prasad and Power, 1997). From an incubation study with a sandy-clay loam soil of *p*H 7.8, Reddy and Prasad (1975) reported an accumulation of 10.5 mg kg^{-1} soil nitrite after one week of incubation and 5.4 mg kg^{-1} soil after two weeks of incubation of urea applied @ of 100 mg kg^{-1} soil.

CROPS SUITABLE FOR ACID AND ALKALINE/ SODIC SOILS

Rice, coffee (*Coffea arabica* L.), pineapple [*Ananas comosus* (L.) Merr.] and tea [*Chamelia sinensis* (L.) Kuntze] grow well on acid soils, while rice, barley (*Hordeum vulgare* L.), sugarbeet (*Beta vulgaris* L.), *dhaincha* [*Sesbania bispinosa* (Jacq.) W. Wight] and karnal grass [*Diplachne fusca* (L.) P. Beauv. ex Roem. & Schutt.] are fairly tolerant to alkaline conditions. Tea is an Al accumulating crop.

SUMMARY

1. Soil water and air together occupy about 50% of soil volume.
2. A proper ratio of soil and water is required for a good plant growth.
3. Soil solution H concentration is measured as *p*H.
4. $pH = -\log_{10} (aH^+)$
5. Exchangeable H is much larger in amounts than H in solution.
6. Exchangeable H is determined for liming acid soils.

7. Soil *p*H determines the concentration and availability of different plant nutrients in soil.
8. Soil *p*H controls the activity of soil organisms.
9. Plants differ in their ability to grow under acid or alkaline conditions in soil.

REFERENCES

Ataur Rahman, M., Barma, N.C.D., Sarker, M.H., Sarker, M.M.R. and Nazrul, M.M.I. 2013. Adaptability of wheat varieties in strongly acidic soils of Sylhet. *Bangladesh Journal of Agricultural Research* **38**(1): 97–104.

Covington, A.K., Bates, R.G. and Durst, R.A. 1985. Definitions of *p*H scales, standard reference values, measurement of *p*H, and related terminology. *Pure and Applied Chemistry* **57**(3): 531–542.

Devi, K.M.D., Gopi, C.S., Santhakumari, G. and Prabhakaran, P.V. 1996. Effect of water management and lime on iron toxicity and yield of paddy. *Journal of Tropical Agriculture* **34**: 44–47.

Foy, C.D. 1992. Soil chemical factors limiting plant growth. *Advances in Soil Science* **19**: 97–149.

Jena, D. and Rout, K.K. 2014. Calcium and magnesium management. (In) *Textbook of Plant Nutrient Management,* pp: 157–181. Prasad R., Kumar D., Rana D.S., Shivay Y.S. and Tewatia R.K. (Eds), Indian Society of Agronomy, New Delhi.

Kumar, B., Dhar, S., Vyas, A.K. and Parmesh, V. 2015. Impact of irrigation schedules and nutrient management on growth, yield and root traits of wheat (*Triticum aestivum*) varieties. *Indian Journal of Agronomy* **60**(1): 87–91.

Linn, D.M. and Doran, J.W. 1984. Effect of water-filled pore space on carbon dioxide and nitrous oxide production in tilled and non-tilled soils. *Soil Science Society of America Journal* **48**: 1,267–1,272.

McLean, E.O., Reicosky, D.C. and Lakshamanan, C. 1965. Aluminum in soils. VII. Interrelationships of organic matter, liming, and extractable aluminum with permanent charge and *p*H dependent cation exchange capacity of surface soils. *Soil Science Society of America Proceedings* **29**: 374–378.

Mukherjee, J.N., Chatterjee, B. and Banerjee, B.M. 1947. Liberation of H^+, Al^{3+} and Fe^{3+} ions from hydrogen clays by neutral salts. *Journal of Colloid Science* **1**: 247–254.

Ponnamperuma, F.N. 1972. The chemistry of submerged soils. *Advances in Agronomy* **24**: 29–96.

Prasad, R. and Power, J.F. 1997. *Soil Fertility Management for Sustainable Agriculture*, 356 pp. CRC Press, Boca Raton, FL, the USA.

Prasad, R., Shivay, Y.S., Kumar, D. and Sharma, S.N. 2006. *Learning by Doing Exercises in Soil Fertility* (A Practical Manual for Soil Fertility), Division of Agronomy, Indian Agricultural Research Institute, New Delhi, 68 pp.

Reddy, R.N.S. and Prasad, R. 1975. Studies on the mineralization of urea, coated urea and nitrification treated urea in soil. *Journal of Soil Science* **26**(3): 304–312.

Rousk, J., Brookes, P.C. and Bååth, E. 2009. Contrasting soil *p*H effects on fungal and bacterial growth suggest functional redundancy in carbon mineralization. *Applied Environmental Microbiology* **75**(6): 1,589–1,596.

Sahrawat, K.L. 1979. Iron toxicity to rice in acid sulfate soils as influenced by water regimes. *Plant and Soil* **51**: 143–144.

Sahrawat, K.L. 2008. Factors affecting nitrification in soils. *Communications in Soil Science and Plant Analysis* **39**(9–10): 1,436-46.

Sanchez, P.A. 1976. *Properties and Management of Soils in the Tropics*, 618 pp. John Wiley, New York, USA.

Singh, G., Singh P.N. and Bhushan, L.S. 1980. Water use and wheat yields in northern India under different irrigation regimes. *Agricultural Water Management* **3**(2): 107–114.

Singh, S.B. and Abrol, I.P. 1988. Long term effects of gypsum application and rice–wheat cropping on changes in soil properties and crop yield. *Journal of the Indian Society of Soil Science* **36**: 316–319.

Singh, V.K., Mishra, R.P., Rani, M. and Singh, S.K. 2014. Water and nutrient interactions in irrigated agriculture. (In) *Textbook of Plant Nutrient Management*, pp. 249–267. Prasad R., Kumar D., Rana D.S., Shivay Y.S. and Tewatia R.K., (Eds), Indian Society of Agronomy, New Delhi.

Zhao, C., Degryse, F., Gupta, V. and McLaughlin, M.J. 2014. Elemental sulfur oxidation in Australian cropping soils. *Soil Science Society of America Journal* **79**(1): 89–96.

QUESTIONS

1. Answer True (T) or False (F).

(a) Rice is an ideal crop for an alkaline soil, but not for an acid soil.
(b) Tea loves soil acidity and is an Al accumulating crop.
(c) Exchangeable-H is an intensity factor.
(d) Excess Al and not the exchangeable-H is the main cause of damage to crops on acid soils.
(e) Soil to solution ratio has no effect in determining soil *p*H.
(f) Sugarbeet is an ideal crop for acid soils.

2. Explain the followings

(a) Soil buffering capacity
(b) Potential vs active acidity
(c) Role of determining exchangeable Al in soils
(d) Available soil moisture
(e) Lime requirement

Max. Marks 100 Sample Test Paper –1 Time 2 Hrs

1. Fill-up the blanks (20 Marks)

(i) Simple Fertilizer Trials on the farmers' fields were started in India at the recommendations of..........
(ii) Phosphorus as an essential plant nutrient was first suggested by ....
(iii) Jethro Tull suggested that .... were the principle of vegetation.
(iv) Essentiality of Zn as a plant nutrient was first suggested by ....
(v) Estimates of C–sequestration by plants at global level are at .... Pg per year.

2. Answer True (T) or False (F) (20 Marks)

(a) CEC of kaolinite is 40 c (+) moles $kg^{-1.}$
(b) Exchangeable-H in soils is a capacity measurement.
(c) Indian Institute of Soil Science is located at Ludhiana.
(d) Sulfur deficiency in plants is expressed in upper (younger) leaves.
(e) Soils having high Smectite minerals are the best suited for tillage operations

3. Explain why? (20 Marks)

(a) 2:1 clay minerals have a net negative charge
(b) Red and yellow soils are poor in soil fertility
(c) Si is considered as a beneficial element
(d) Ni is considered as an essential element
(e) C-sequestration by soil is considered important for the environment

4. Briefly discuss the followings (40 Marks)

(a) Variable charge in soil
(b) Soil *p*H and availability of essential plant nutrients
(c) Soil C budget
(d) Soil porosity
(e) Micronutrients

Chapter 6

Soil Nitrogen and Nitrogen Cycle

Introduction

Nitrogen constitutes 78% of the atmospheric gases and there is about 75,000 Mg N on each hectare of land. However, 99% of the atmospheric N is as N_2 and is inert and non-reactive, while the rest 1% or even less is the reactive-N. Reactive-N is linked with H, O and C and is responsible for the creation and sustenance of all life on the planet Earth. Nitrogen is the key nutrient that separates proteins from carbohydrates and fats, the three major macronutrients for humans. As a component of proteins, N is present in DNA and RNA, which carry the hereditary characters and then as the major constitute of the body muscles of humans (and animals). Plant growth and crop production are also very much dependent on the adequate availability of N in soil.

SOIL NITROGEN

Global amounts of total nitrogen in the upper 1 m layer soil is estimated at 133–140 Pg N (Batjes, 2014). Nitrogen in soil is present in organic and inorganic form.

Organic N

This constitutes 95–99% of total soil N. Total N in Indian soils varies from 0.02 to 0.1% (0.2 to 1 g kg^{-1} soil) in most cultivated soils; the values being a little higher (0.32% or 3.2 g kg^{-1} soil) in hill soils (Kanwar and Takkar, 1963). As a contrast total soil N is much higher in temperate regions. Soil organic N is generally present in the surface 0–20 cm soil layer. Soil organic N is mostly proteins, but some other N-compounds, such as glucosamine, free amino acids are also present.

Inorganic N

Inorganic N in soils is present as ammonium, nitrate, exchangeable ammonium and fixed ammonium. In most well drained soils N is mostly present as nitrate-N and in Indian soils it could vary from 3 to 36 mg kg^{-1} (Shukla and Singh, 1968). The values for ammonium-N could be about 2–20 mg kg^{-1}. The amount of exchangeable ammonium would depend on soil texture and clay minerals present and Sen *et al.* (1957) reported values of 10–50 mg kg^{-1}.

Fixed ammonium N

Fixed ammonium in Indian soils varies from 70 to 798 mg kg^{-1} and in soils with high 2 : 1 clay minerals it may account 5–10% of total N in surface soil and 20–40% of total N in sub-soil (Grewal and Kanwar, 1967).

SOIL NITROGEN DYNAMICS

Once the plant and animal residues are applied or reach the soil on their own, they are first broken down to fine particles by different macro (earthworms, termites etc.) and meso-fauna (mites, *Colembolla* etc.). The heterotrophic microfauna (bacteria and fungi) further breakdown the N compounds (proteins, amino acids, amino sugars etc.) to ammonium (NH_4^+). This microbial process is known as *ammonification*. Ammonium so formed may be partly lost by volatilization specially on calcareous high *p*H soils, be taken up by bacteria or plants for their body built up (*immobilization*), be adsorbed in the exchange complex of clay minerals or be fixed in the crystal lattice of 2 : 1 silicate minerals to form micas. Most plants take up N as nitrate and ammonium both. However, some plants, such as rice, are ammoniphilic and may prefer ammonium (Prasad *et al.*, 1983). Ammonium (NH_4^+) is converted to nitrite (NO_2^{-1}) by autotrophic bacteria *Nitrosomonas* sp. and *Nitrosococcus* sp. (Step I) and then NO_2 is converted to NO_3 by autotrophic bacteria *Nitrobacter* sp. Nitrification is brisk at *p*H 5.4 to 7.8 (Morrill and Dawson, 1967), –33 KPa (near field capacity) moisture tension (Malhi and McGill, 1982) and at temperatures 20–37°C. Nitrification is slow at *p*H values less than 5, temperatures below 20°C and moisture tensions below or above field capacity. Of the two steps of nitrification, step II is faster than step I, so nitrites are in soil only for a short while. However, under high *p*H conditions (above *p*H 8.0), step II could be slower and nitrites may accumulate (Prasad and Power, 1997).

Nitrates formed are taken up by plants or may be leached down the profile or may be denitrified to nitrous oxide (N_2O), other N-oxides or even nitrogen gas (N_2). These gases are lost to the atmosphere. This series of processes is a major component of N-cycle in nature (Fig. 6.1).

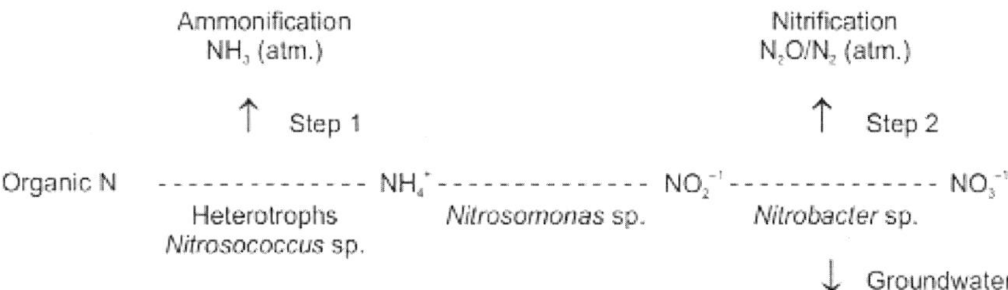

Fig. 6.1: Soil nitrogen dynamics

GAINS AND LOSSES OF SOIL NITROGEN

Gains and losses of soil N are presented in Fig. 6.2 and are briefly discussed.

Gains	**Losses**
Manures	Removal by crops
Fertilizers	Soil erosion
Rhizobial N fixation	Surface runoff
N–fixation by BGA	Ammonia volatilization
N–fixation by *Azolla–Anabaena*	Denitrification
N–fixation by *Azotobacter* and *Azospirillum*	Leaching
Addition as NO_3 through rainwater	

Fig. 6.2: Gains and losses of soil nitrogen

Soil nitrogen gains

Manures and composts: Farmyard manure (FYM) generally supplies 4–5 kg N t^{-1}, while composts especially vermicompost can supply 10–15 kg N t^{-1}. Nevertheless cost of one unit of N is much higher in manures and composts and farmers are generally ignoring them. Manures, especially composts are now mostly used in organic agriculture.

Fertilizers: Fertilizer is the main source of N in India, especially after the introduction of high-yielding varieties of rice, wheat and Bt cotton, where applications per crop season vary from 100 to 150 kg N ha^{-1}. Even dryland crops receive 30–60 kg N ha^{-1}.

Table 6.1: Amount of nitrogen-fixed by some legumes

Legume	N fixed kg ha^{-1}
Chickpea	103–120
Cowpea	73–354
Lentil	88–114
Mungbean or green gram	63–342
Fieldpea	52–77
Urdbean or blackgram	119–140
Alfalfa	229–290

Source: Vyas *et al*. (2014)

Rhizobial N fixation: Legumes as soil fertility-restoring crops have been known since ages (Buttar and Rana, 2014). Growing of legumes for green manuring or as a short-duration catch crop, such as, summer mungbean [*Vigna radiata* (L.) R. Wilczet] has received considerable attention in the rice–wheat cropping system of north India (Sharma *et al.*, 2010). Legumes generally fix 20–25 kg N per tonne of biomass produced (Unkovitch *et al.*, 2010). For this reason forage legumes, such as, Egyptian clover (*Trifolium alexandrinum* L.) fix much more N. Some data are in Table 6.1. Fertilizer N equivalent for the residual effect of growing a summer mungbean crop and incorporating its residue could be 50–60 kg N ha^{-1} or even more (Sharma and Prasad, 1999).

N fixation by BGA: In rice field blue green algae (BGA) can fix 25–30 kg N ha^{-1} (Prasanna *et al.*, 2014)

Azolla–Anabaena association can fix 40–60 kg N ha^{-1} in rice fields (Prasanna *et al.*, 2014). Use of*Azolla–Anabaena* association is quite popular in Vietnam.

Azotobacter and Azospirillum: These organisms can fix about 15–30 kg N ha^{-1} (Prasanna *et al.*, 2014).

Although *Rhizobial* N-fixation has received much attention, Ladha *et al.* (2016) observed that 370 Tg N (24% of the total 1,551 Tg of N removed globally by three cereals, namely, rice, maize, wheat in the world) could be attributed to N-fixation by non-symbiotic N-fixation; this amounted to 22 kg N ha^{-1} for rice and 13 kg N ha^{-1} for maize and wheat. Higher values for rice could be owing to higher N fixation by BGA and *Azolla*, which are associated with the rice culture.

Addition as nitrates through rainwater: About 8–10 kg N ha^{-1} could be added by rains in a year.

Soil nitrogen losses

Erosion and run-off: In a 5-year (2005–2009) study at Hidayatnagar research farm of the ICAR-Central Research Institute for Dryland Agriculture, Hyderabad, it was found that under semi-arid tropical conditions, the average nutrient loss due to erosion and surface runoff was 20.6 kg C, 5.05 kg N, 0.05 kg P and 9.94 kg K ha^{-1} $year^{-1}$ (Mandal *et al.*, 2012). From China, Yen *et al.* (2005) reported a run-off loss of 6 kg N ha^{-1} $year^{-1}$ from Baiyandigan lake region and 14.8 kg N ha^{-1} $year^{-1}$ from Taiku lake basin of China. The nutrient losses due to erosion and surface runoff are likely be much higher under high rainfall conditions and more sloping lands.

Ammonia volatilization: Urea hydrolysis is fairly fast under Indian condition and bulk of urea hydrolyses take place within two weeks of application on a neutral or alkaline soil (Reddy and Prasad, 1975) and the resulting *p*H is generally above 8.5. This leads to NH_3 volatilization loss. Pathak *et al.* (2006) based on simulation studies reported an ammonia volatilization loss of 16–62 kg N ha^{-1} with an average of 30 kg N ha^{-1} in the rice–wheat cropping system belt of the Indo-Gangetic belt of north India. Yan *et al.* (2015) reported an ammonia volatilization loss of 11.6 to 57 kg N ha^{-1} from a 3-year study in rice field in Zhejiang Province, China.

Leaching: Leaching studies in fields are difficult. In a pot-culture study with rice (Prakasa Rao and Prasad, 1980) found that, N-leaching loss was 64.5% with sodium nitrate, 18.5% with calcium ammonium nitrate, 13.7% with ammonium sulphate, 11.5% with urea and 9.5% with neem-coated urea. Thus, leaching loss of N is most with nitrates. Vinod *et al.* (2015) reported that in Karnataka, the average NO_3-N leached from irrigated areas varied from 51.2 to 74.9 kg N ha^{-1}. Pathak *et al.* (2004) observed that, N loss from leaching in the rice–wheat cropping system of the Indo-Gangetic belt of north India could be 10% of applied N. Bharadwaj *et al.* (2012) reported that, 28% of shallow waters in Ludhiana district contained 11–25 mg NO_3-N L^{-1}, which is above the US-Environmental Protection Authority (EPA) safe limit of 10 mg NO_3-N L^{-1}.

Denitrification: Denitrification loss of nitrogen as N_2O from soil and water has received more attention from the environmentalists than agricultural scientists, because it is useful in reducing inland and marine pollution of nitrate-N. Denitrification as per definition is a biological process, however, some of it may take place by chemical reactions, such as, conversion of Fe^{2+} to Fe^{3+} (Moraghan and Buresh, 1977). Iron-containing minerals siderite (Rakshit *et al.*, 2008) and mangnetite (Dhakal *et al.*, 2013) can also reduce NO_3^-. As regards, biological denitrification, bacteria as well as fungi are involved in the production of N_2O from nitrates. Again, the bacteria could be heterotrophs, such as, *Bacillus* sp., *Agrobacterium* sp. etc or autotrophs, such as, SF-oxidizing *Thiobacillus* sp. etc. (Tiedje, 1988). Similarly, there is a large group of fungi, such as, *Alternaria alternata*, *Aspergillus flavus* etc. that can reduce nitrate to nitrite (Mothapo *et al.*, 2015). In addition, some N_2O is produced by dissimilatory nitrate reduction to ammonium (DNRA) by some bacteria and fungi (Kasper, 1982), although its overall contribution to total N_2O production from soil could be little, say about 3% (Silver *et al.*, 2005). Thus, there are many pathways for reduction of nitrates and nitrites to N_2O.

In India, Aulakh *et al.* (2001) reported that 22–33% of N applied to rice could be lost by denitrification, while Pathak *et al.* (2004) reported that average N loss by denitrification was 10–15 kg N ha^{-1} in rice and 5–10 kg N ha^{-1} in wheat in the rice–wheat cropping system. These values are determined by subtraction method, where unaccounted losses are taken as loss by denitrification. Actual recovery of (N_2 + N_2O)-^{15}N values reported are quite low, for example, 0.5% of the applied N by John *et al.* (1989) suggesting the need for improvement in field-measurement techniques. Nitrogen losses are the most under alternate submergence and drainage conditions (Prasad and Rajale, 1972); such as those obtained under irrigated rice in the rice–wheat cropping system of north India (Prasad, 2005).

SUMMARY

1. Nitrogen constitutes 78% of the atmospheric gases, but 99% of the atmospheric N is as N_2 and is inert and non-reactive, while the rest 1% or even less is the reactive N.
2. Nitrogen in soil is present in organic and inorganic form. Organic N constitutes 95–99% of total soil N. Total N in Indian soils varies from 0.02 to 0.1% (0.2 to 1 g kg^{-1} soil) in most cultivated soils.
3. Inorganic N in soils is present as ammonium, nitrate, exchangeable ammonium and fixed ammonium. In most well-drained soils in India, nitrate-N could vary from 3 to 36 mg kg^{-1}, while the values for ammonium-N could be about 2–20 mg kg^{-1}.
4. Fixed ammonium in Indian soils varies from 70 to 798 mg kg^{-1} and in soils with high 2 : 1 clay minerals it may account 5–10% of total N in surface soil.
5. Soil organic N is first acted upon by heterotrophs, which produce NH_4^+, which is then oxidized to nitrite (NO_2) by *Nitrosomonas* sp. and *Nitrosococcus* sp. (Step I); nitrite is then oxidized to nitrate (NO_3^{-1}) by *Nitrobacter* sp. (Step II), and the process is known as nitrification.
6. Soil N gains include manures, fertilizers, N–fixation by *Rhizobia*, BGA, *Azolla-Anabaena*, *Azotobacter* and *Azospirillum* and NO_3 through rain water.
7. Soil N losses include removal by crops, erosion and surface runoff, ammonia volatilization, denitrification and leaching.
8. Legumes can fix about 50–300 kg N ha^{-1}, while *Azotobacter* and *Azospirillum* can fix about 15–30 kg N ha^{-1}.
9. Ammonia volatilization losses and denitrification losses in the rice–wheat cropping system could be about 30 kg N ha^{-1} each per cycle of the cropping system.
10. Denitrification is carried out by bacteria as well as fungi.

REFERENCES

Aulakh, M.S., Khera, T.S., Doran, J.W. and Bronson, K.F. 2001. Denitrification N_2O and CO_2 fluxes in rice–wheat cropping system as affected by crop residues. *Biology and Fertility of Soils* **34**(6): 375–389.

Batjes, N.H. 2014. Total carbon and nitrogen content in the soils of the world. *European Journal of Soil Science* **65**: 10–21.

Bhardwaj, A.S., Garg, S., Sondhi, S.K. and Taneja, D.S. 2012. Nitrate contamination of shallow aquifer groundwater in the central districts of Punjab. *Journal of Environment Science and Engineering* **54**(1): 90–97.

Buttar, G.S. and Rana, D.S. 2014. Legumes as green manures. (In) *Textbook of Plant Nutrient Management*, pp. 311–324. Prasad, R., Kumar, D., Rana, D.S., Shivay, Y.S. and Tewatia, R.K. (Eds). Indian Society of Agronomy, New Delhi.

Dhakal, P., Matocha, C., Huggins, F. and Vandiviere, M. 2013. Nitrite activity with nitrite. *Environmental Science and Technology* **47**: 6,206–6,213.

Grewal, J.S. and Kanwar J.S. 1967. Ammonium fixation in Punjab soils. *Journal of Research, Punjab Agricultural University*, Ludhiana **4**: 1–6.

John, P.S., Buresh, R.J., Prasad, R. and Pandey, R.K. 1989. Nitrogen gas (N + N_2O) flux from urea applied to lowland rice as affected by green manure. *Plant and Soil* **119**: 7–13.

Kanwar, J.S and Takkar, P.N. 1963. Sulphur, nitrogen and phosphorus deficiency in tea soils of Punjab. *Indian Journal of Agricultural Sciences* **33**: 291–294.

Kasper, H. 1982. Nitrate reduction to nitrous oxide by propionibacteria: Detoxification mechanisms. *Archives of Microbiology* **133**: 126–130.

Ladha, J.K., Padre, A.T., Reddy, C.K., Cassman, K.G., Verma, S., Powlson, D.S., van Kessel, C., Richter, D.B., Chakraborty, D. and Pathak, H. 2016. Global nitrogen budgets in cereals: A 50 year assessment for maize, rice and wheat production systems. *Scientific Reports* **6**: 193555 (doi:10 .1038/srep193555).

Malhi, S.S. and McGill, W.B. 1982. Nitrification in three Alberta soils: effect of temperature, moisture and substrate concentration. *Soil Biology and Biochemistry* **14**: 393–399.

Mandal, U.K., Sharma, K.L., Prasad, J.V.N.S., Sanjeeva Reddy, B., Narsimlu, B., Saikia, U.S., Adeka, R.V., Yadaiah, P., Masane, R.N., Masane, R.N, Venkatravamma, K., Satyam, B., Raju, B. and Srivastava, N.U. 2012. Nutrient losses by runoff and sediment from an agricultural field in semi-arid tropical India. *Indian Journal of Dryland Agriculture Research and Development* **27**(1): 1–9.

Moraghan, J. and Buresh, R.J. 1977. Chemical reduction of nitrite and nitrous oxide by ferrous iron. *Soil Science Society of America Journal* **41**: 47–49.

Morril, L.G. and Dawson, J.E. 1967. Patterns observed for the oxidation of ammonium to nitrate by soil organisms. *Soil Science Society of America Journal* **31**: 757–760.

Mothapo, N., Chen, H., Cubeta, M.A., Crossman, J.M., Fuller, F. and Shi, W. 2015. Phylogenic, taxonomic and functional diversity of fungal denitrifiers and associated N_2O production efficacy. *Soil Biology and Biochemistry* **83**: 160–175.

Pathak, H., Li, C., Wasserman, R. and Ladha, J.K. 2006. Simulation of nitrogen balance in rice–wheat system for the Indo-Gangetic plain. *Soil Science Society of America Journal* **70**: 1,612–1,622.

Pathak, H., Tismina, J., Humphreys, E. and Godwin, D. 2004. Simulation of rice crop performance and water and nitrogen dynamics, and emission in northwest India using CERES Rice model. *Technical Report CSIRO Land and Water* **23**(4): 111.

Prakasa Rao, E.V.S.P. and Prasad, R. 1980. Nitrogen leaching losses from conventional and new nitrogenous fertilizers in lowland rice culture. *Plant and Soil* **57**: 383–392.

Prasad, R. 2005. Rice–wheat cropping systems. *Advances in Agronomy* **86**: 255–339.

Prasad, R. and Power, J.F. 1997. *Soil Fertility Management for Sustainable Agriculture*, 356 pp. CRC Press, Boca Raton, FL, the USA.

Prasad, R. and Rajale, G.B. 1972. The influence of nitrification inhibitors and slow-release nitrogen materials on transformation of fertilizer nitrogen in soils of fluctuating moisture content. *Soil Biology and Biochemistry* **4**: 451–457.

Prasad, R., Thomas, J. and Gupta, V.V.S.R. 1983. Ammoniphillic plants for reducing water pollution. *Environmental Conservation* **10**: 260–261.

Prasanna, R., Wattal-Dhar, D. and Shivay, Y.S. 2014. (In) *Textbook of Plant Nutrient Management,* pp: 231–248. Prasad, R., Kumar D., Rana, D.S., Shivay, Y.S. and Tewatia, R.K. (Eds). Indian Society of Agronomy, New Delhi.

Rakshit, S., Matocha, C. and Coyne, M. 2008. Nitrite reduction by siderite. *Soil Science Society of America Journal* **72**: 1070–1077.

Reddy, R.N.S. and Prasad, R. 1975. Studies on the mineralization of urea, coated urea and nitrification inhibitor treated urea in soil. *Journal of Soil Science* **26**: 304–312.

Sen, A., Rewari, R.B. and Pandey, S.L. 1957. Ammonium content of some Indian soils in relation to their reaction and hygroscopic moisture. *Journal of The Indian Society of Soil Science* **5**: 109–112.

Sharma, S.N. and Prasad, R. 1999. Effects of *Sesbania* green manuring and mungbean residue incorporation on productivity and nitrogen uptake of a rice–wheat cropping system. *Bioresource Technology* **67**: 171–175.

Sharma, S.N., Prasad, R., Dwivedi, M.K., Kumar, S., Davari, M.R. and Shukla, L. 2010. Crop diversification and residue incorporation for making rice–wheat cropping systems sustainable. *Journal of Sustainable Agriculture* **34**: 342–364.

Shukla, G.C. and Singh, M. 1968. A new photoelectric colorimeter for determining nitrate in soils. *Journal of the Indian Society of Soil Science* **16**: 77–82.

Silver, W.L., Thompson, A.W., Reich, A., Ewel, J.J. and Firestone, M.K. 2005. USDA Forest Service, Pacific Southwest Research Station, Institute of Pacific Island Forestry, 1151 Punchbowl Street Room 323, Honolulu, Hawaii 96813 USA. Nitrogen cycling Ecosystem Sciences Division, Department of Environmental Science, Policy, and Management, 137 Mulford Hall no. 3114, University of California, Berkeley, California 94720 USA in tropical plantation forests: Potential control of nitrogen retention. *Ecological Applications* **15**: 1,604–1,614.

Tiedje, J.M. 1988. Ecology of denitrification and dissimilatory nitrate reduction to ammonium. (In) *Biology of Anaerobic Organisms,* pp: 179–243. Zehnder, A.J.B. (Ed), John Wiley, New York, USA.

Unkovitch, M.J., Baldock, J. and Peoples, M.B. 2010. Prospects of simple linear models for estimating symbiotic N_2 fixation by crop and pasture legumes. *Plant and Soil* **329**: 75–89.

Vinod, P.N., Chandramouli, P.N. and Koch, M. 2015. Estimates of nitrate leaching in groundwater in agriculturally used urea in the state, Karnataka, India using existing models and GIS. (In) *Proceedings of the International Conference on Water Resource, Coastal and Ocean Engineering*, 12–14 March, 2015 at National Institute of Technology, Survatha, Karnataka.

Vyas, A.K., Kumar, D., Ali, M., Prasad, R. and Rana, D.S. 2014. Legumes in crop production systems. (In) *Textbook of Plant Nutrient Management,* pp: 295–310. Prasad, R., Kumar, D., Rana, D.S., Shivay, Y.S. and Tewatia, R.K. (Eds), Indian Society of Agronomy, New Delhi.

Yan, L., Chen, Y., Wu, C.Y., Xu, T. and Ji, X.J. 2015. Determination of optimum nitrogen application rates in Zhejiang Province, China, based on rice yields and ecological security. *Journal of Integrative Agriculture* **14**(12): 2,426–2,433.

Yen, W., Zhang, W., Chen, X. and Tang, Y. 2005. Nitrogen export by runoff from agricultural plots in two basins of China. *Nutrient Cycling in Agroecosystems* **71**(2): 121–129.

QUESTIONS

1. Answer True (T) or False (F)

(a) Some chemical reactions in soil can also produce nitrous oxide.
(b) There is not much fixed ammonium-N in Indian soils.
(c) Reactive-N is about 10% of the total N in the atmosphere.
(d) Denitrification is carried out by bacteria and fungi have no role in it.
(e) Growing of summer mungbean in the rice–wheat cropping system can contribute about 60 kg N ha^{-1}.

2. Discuss briefly

(a) Ammonia volatilization.
(b) N-fixation by legumes
(c) Nitrogen dynamics in soil.

3. Match the followings

(a)	Tiedje	(i)	Nitrogen losses from rice–wheat cropping system
(b)	Kasper	(ii)	Chemical reduction of nitrate
(c)	Moraghan	(iii)	Dissimilatory nitrate reduction to ammonium (DNRA)
(d)	Pathak	(iv)	Nitrate content in shallow wells in Ludhiana district.
(e)	Bharadwaj	(v)	Nitrate reduction by heterotrophic and autotrophic bacteria.

Chapter 7

Fertilizer Nitrogen Management and its Environmental Implications

Fertilizer N has been the key input in increasing cereal production throughout the world. In India, it made possible to achieve the yield potential of high-yielding dwarf varieties of wheat that led to 'Green Revolution in 1968'. Nitrogen fertilization also increases protein content in wheat grains (Rajeswara Rao and Prasad, 1987) and leads to increased production of proteins in cereals (Prasad and Shivay, 2015). Cereal production in India increased from 42.4 million tonnes (Mt) in 1950–51 to 259.6 Mt in 2017–18, while fertilizer N application increased from 0.05 Mt to 16.96 Mt. However, the average agronomic efficiency of N (AEn) in India for cereals is only 9.5 kg grain kg^{-1} N (Sharma and Batra, 2011) as compared to global average of 24 kg grain kg^{-1} N for maize, 22 kg grain kg^{-1} N for rice and 18 kg grain kg^{-1} N for wheat (Ladha *et al.*, 2005). Besides this, more disturbing is the fact that in irrigated areas the AEn declined from 13.4 to 3.4 kg grain kg^{-1} N during 1970–2005 (Biswas and Sharma, 2006). This is due to poor N management. Bulk of fertilizer N is applied broadcast on surface in most crops.

NITROGEN USE EFFICIENCY

Nitrogen-use efficiency (NUE) means different things to different people. Some of the most commonly used terms related to NUE are given below (Prasad, 2009).

Agronomic efficiency (AEn) = (Yf–Yc)/N = $\Delta Y / \Delta N$ (kg grain/kg N applied)
Apparent recovery efficiency (AREn) = 100 (Uf – Uc)/ N = $\Delta U / \Delta N$ (% of applied N)

Physiological efficiency (PEn) = (Yf – Yc)/ Uf–Uc = $\Delta Y / \Delta U$ = AE/RE (kg grain/kg N absorbed)

Partial factor productivity of fertilizer (PFPn) = Y/N (kg grain/kg N applied)

In the above expressions, Yf and Yc are grain (or other economic produce) yield in kg ha^{-1} in fertilized and the control (check) plots, while Uf and Uc refer to the nitrogen uptake (kg ha^{-1}) in fertilized and control plot, respectively, and N refers to the nitrogen applied (kg ha^{-1}). The PFPn does not take into account the control plot yield, but is an useful term because it can be calculated for the experiments on tillage, sowing dates, seed rates, weed control, water management etc, where a blanket fertilizer application is made. The PFPn can also be used for comparing NUE in different countries or in different parts of a country. When tracers such as ^{15}N are used, the recovery is known as true recovery efficiency (TREn). A lower case letter as suffix is generally provided to indicate the nutrient for which the nutrient-use efficiency (NUE) refers to, for example, AEn refers to the Agronomic efficiency of N.

Globally the apparent recovery efficiency (AREn) in experimental plots was found to be 54% in wheat and 44% in rice (Doberman, 2005). From India, Jat *et al.* (2013) based on studies made at 22 centres reported AREn value ranging from 36.3 to 79.4% in wheat. True recovery efficiency of N (TREn) is generally lesser than AREn and generally decreases with the rate of application. In rice, George and Prasad (1989) reported a TREn value of 31.0, 26.7 and 25.9% for applications of 50, 100 and 150 kg N ha^{-1}, while Bandyopadhyay and Sarkar (2005) reported a TREn value of 28.8% for rice with an application of 180 kg N ha^{-1}. Xiao-Hui *et al.* (2006) reported TREn values of 13.6 and 28.1% for single and split applications of 135–270 kg N ha^{-1}, respectively for rice in China. Again from China, Chen *et al.* (2016) reported TREn values of 18.0 and 31.7% for single and split applications of 240 kg N ha^{-1}, respectively for wheat in China.

Prasad *et al.* (2000) reported a physiological efficiency of N (PEn) values of 40–44 kg grain kg^{-1} for rice and 24–48 kg grain kg^{-1} for wheat. In the same study, the partial factor productivity of N (PFPn) values decreased from 84 kg grain kg^{-1} N for both rice and wheat at 40–60 kg N ha^{-1} to 32 kg grain kg^{-1} N at 121–180 kg N ha^{-1}. In general, use efficiency and PFP values are lower at high N rates.

It needs to be brought out that since crop yield is the nominator for all NUE terms, the values will increase with an increase in crop yield. Therefore, all factors affecting crop yield will affect NUE. These include: variety, seed quality, timely sowing/ transplanting, use of organic manure, adequate supply of all nutrients other than N, efficient water management, efficient weed management, adequate pest management and of course N management, which are discussed below.

NITROGEN MANAGEMENT

Nitrogen management has to be based on the 4R nutrient stewardship concept (Bruulsema *et al.*, 2012), which involves right source at the right rate, time and place.

Sources of nitrogen

At present four straight N (containing only N), five NP and six NPK fertilizers supplying N are manufactured in India (Table 7.1). The total amount of fertilizer manufactured in India during 2018–19 was 41.59 million tonnes (nitrogenous 24.61 + phosphate 4.1 + NP/NPK complex 12.88 million tonnes). Urea (*now* neem-coated urea or NCU) alone meets about 80% fertilizer need of the country.

Table 7.1: Nitrogen fertilizers manufactured in India in the year 2018–19

Fertilizer	Nutrient content (% N-% P_2O_5–% K_2O)	Amounts manufactured in India (Million tonnes)	Million tonnes (N-P_2O_5–K_2O)
Straight N fertilizers			
Urea	46–0–0	24.61	11.14 (N)
Calcium ammonium nitrate (CAN)	25–0–0		
Ammonium sulphate (AS)	20.6–0–0		
Ammonium chloride (ACl)	25–0–0		
NP fertilizers			
Diammonium phosphate (DAP)	18–46–0	8.6	2.19 (N)+3.94 (P_2O_5)
Monoammonium phosphate (MAP)	11–52–0		
Urea ammonium phosphate (UAP)	24–24–0		
Urea ammonium phosphate (UAP)	28–28–0		
Ammonium phosphate sulphate (APS)	16–20–0		
Ammonium phosphate sulphate (APS)	20–20–0		
Nitrophosphate (NP)	20–20–0		
Nitrophosphate (NP)	23–23–0		
NPK fertilizers			
Nitrophosphate K (NP-K)	15–15–15		
NPK	17–17–17	4.27	–
NPK	19–19–19		
NPK	14–28–14		
NPK	12–32–16		
NPK	10–26–26		
NPK	14–35–14		

Considerable research has been done in India on nitrification inhibitors blended N fertilizers (Prasad and Power, 1995) and on slow-release N fertilizers including indigenous materials, such as, neem-coated urea (Prasad and Shivay, 2015; Prakasa Rao *et al.*, 2016), sulphur-coated urea (Shivay *et al.*, 2016b) and the outcome of this research was the development of neem-coated urea (NCU). All urea manufactured in India is now coated with neem oil.

For basal application (at sowing) for all crops, both under dryland agriculture conditions or under irrigation, NP and NPK fertilizers are the best, the level of N and K is decided by the rate at which P is applied. Additional N, if needed can be added as NCU or any other N fertilizer.

Rate of application

Current recommendations for N rates to be applied to different crops are generally based on the studies on crop responses to N obtained from agronomic field experiments. Soil tests for available N, viz., SOC or alkaline permanganate-oxidisable N, has not been of much use. The newly developed 'need-based nitrogen management' involving measurement of the chlorophyll of the leaves using leaf-colour chart or SPAD meter (Varinder-pal Singh *et al.*, 2010) holds some promise for deciding upon top-dressing to be made. This technique along with foliar application can save some nitrogen without sacrificing yield.

Time of application

In dryland agriculture, all N is applied at sowing, while in irrigated agriculture it is applied in 2 or 3 split doses. Application in two doses, half dose at sowing and the rest half at first irrigation in wheat, at panicle initiation/ maximum tillering in rice or at active vegetative growth stage in other crops is the general recommendation as well as the common practice. In hybrid maize, the general recommendation is application in three equal (or some minor changes) doses at planting, knee-high stage and tasseling. In sugarcane, N is generally applied in three equal doses at planting, active tillering and after rains. Recommendations for sugarcane vary considerably depending on the region and wheather the crop is *eksali* (one year) or *adsali* (18 months).

Method of application

Basal dose of N along with all P and K at sowing in most crops is deep placed in soil using a fertilizer drill drawn by a tractor or by a *pora* (a metallic funnel) attached to a *desi* plough, both under dryland agriculture conditions or under irrigation. In transplanted rice, basal fertilizer is broadcast and incorporated in soil at final puddling. Top-dressing with N is mostly done by broadcast method, except in widely spaced row crops, such as maize, cotton, sugarcane, where it is banded by hand. Foliar application, which is an efficient method of N application, has not received much attention in India. Foliar-applied N is rapidly absorbed by the plants and there are no losses as from the soil application. Mostly urea is used in foliar spray of N, because it is non-polar and does not adversely affect the foliage; the safe limit is 3% urea solution using a knap-sack sprayer. Further, urea goes well with most herbicides, pesticides and micronutrients. Smith *et al.* (1991) using ^{15}N labelled urea found that when applied to foliage at heading in wheat, 69% of the applied N was in the above-ground portion of the plant. Bhuiyan *et al.* (2012) from Bangladesh reported that, AEn of foliar applied through urea in rice was 93.8 kg grain kg^{-1} N as compared to 43.78 kg grain kg^{-1} N for soil-applied urea.

However, using the knap-sack sprayers, generally available with farmers only 10 kg N ha^{-1} or less can be applied in one application (Prasad and Shivay, 2016). Nevertheless, urea should be generally included whenever herbicide/ pesticide or a micronutrient is foliar applied to a crop. Data on the amount of N economy possible through one or two foliar sprays of urea in different crops need to be generated.

INTERACTION WITH OTHER NUTRIENTS

Two nutrients are said to interact, when the application of one nutrient positively or negatively influences the effect of the other nutrient. Fortunately N interacts positively with most other nutrients. To explain this some data are provided in Table 7.2, which show that AEn (kg grain increased kg^{-1} N) of N increased from 3.83 without P or K to 4.9 when K was applied along with N, and to 14.33, when P was also applied along with N and finally it increased to 18.25, when both P and K were applied with N. This shows that both P and K interacted positively with N. These data also bring out the importance of balanced NPK fertilization for rice. Thousands of such data exist from all over India to show the need for adequate P and K fertilization to increase the efficiency of applied fertilizer N (Sharma and Batra, 2011)

Table 7.2: Response of rice to NPK fertilisation and agronomic efficiency of nitrogen (AEn) in presence of other nutrients

Nutrient (kg ha^{-1})	Grain yield (kg ha^{-1})	Response to nutrient	AEn (kg grain increased kg^{-1} N)
Control (no nutrient)	3,030	–	–
120 N	3,490	460	3.83
120 N + 40 K	3,620	590	4.92
120 N + 40 P	4,750	1,720	14.33
120 N + 40 P + 40 K	5,220	2,190	18.25

Source: Paul *et al.* (2013)

Table 7.3: Effect of nitrogen fertilization on the concentration (mg kg^{-1} grain) of iron, zinc, manganese and copper in wheat grain

kg N ha^{-1}	Fe	Zn	Mn	Cu
0	150	37.3	41.0	5.1
130 as urea	156	39.2	43.7	5.8
130 as 5% sulphur-coated urea	181	44.5	46.8	7.7
LSD (P=0.05)	4.7	1.9	2.3	0.4

Source: Shivay *et al.* (2016a)

Brennan and Bolland (2008) reported a significant positive N × S interaction in canola from southwestern Australia from 17 field experiments conducted during 1994-2005. In all experiments the grain yield response to N increased significantly with the increase in S application. Also the grain yield response to S increased as the level of N was increased.

A positive interaction between N (applied as urea or sulphur-coated urea) and micronutrients Fe, Zn, Mn and Cu from the viewpoint of their concentration in wheat grain was reported by Shivay *et al.* (2016a) (Table 7.3); the values were higher with sulphur-coated urea, showing a positive interaction with S application on the concentration of these micronutrients in wheat grain.

ENVIRONMENTAL IMPLICATIONS

As already discussed, about 40–50% of applied fertilizer N is taken up by the crops, some may be immobilized by microorganisms in the soil but the rest is lost by erosion, surface runoff, leaching and denitrification. These channels of N loss create environmental problems.

Nitrogen lost by erosion and surface runoff leads to eutrophication of inland lakes in China (Le *et al.*, 2010) and elsewhere, and marine waters in Gulf of Mexico (Turner *et al.*, 2008) and even in Bay of Bengal (Chowdhury, 1989). Nitrogen eutrophication of marine waters has already reached alarming levels from the viewpoint of marine life, resulting in the mortality of fish and other sea organisms (Vitousek *et al.*, 2009).

Leaching of nitrate leads to nitrate pollution of groundwater, the major drinking water source all over the world. Nitrate content in drinking water above 10 mg NO_3-N L^{-1} (45 mg nitrate L^{-1}) (safe limit as per USEPA) leads to methaemoglobinemia or blue baby syndrome (Knobleboch *et al.*, 2000). Bharadwaj *et al.* (2012) has reported that 29% of shallow well waters in Ludhiana district in Punjab have nitrate content above the safe level. Zhou *et al.* (2016) reported that in northern China, the nitrate-N level in drinking waters in vegetable-growing regions was 70 mg NO_3–N L^{-1} or even more.

As regard denitrification, it is estimated that fertilizer N may contribute up to 6.36% of the total anthropogenic contribution of nitrous oxide (Smith, 2010), which was 12.7 Tg N_2O–N year^{-1} in 1990 and is estimated to be 25.7 Tg N_2O–N year^{-1} in 2100 (Kroetz, 1994). The N_2O emission in India was only 0.25 Tg N_2O–N year^{-1} in 1995 and is estimated to increase to 0.84 Tg N_2O–N year^{-1} in 2035 (Garg *et al.*, 2003). It may be mentioned that global warming effect of N_2O is ~300 times that of CO_2. In addition to global warming, N_2O is also detrimental to ozone layer (Ravishankara *et al.*, 2009).

Thus in addition to monetary loss to the farmers and the country, poor N management also creates environmental problems.

SUMMARY

1. Nitrogen-use efficiency (NUE) means different things to different people. Agronomic efficiency of N (AEn) is kg grain kg^{-1} N over the control; Apparent recovery efficiency of N (AREn) is percentage of applied N taken up by crop; Physiological efficiency of N (PEn) is kg grain kg^{-1} N taken up by crop and Partial Factor Productivity of N (PFPn) is kg grain kg^{-1} N without taking the control plot yield into consideration.
2. The NUE is fairly low in India. Agronomy efficiency of nitrogen (AEn) in India for cereals is ~ 9.5 kg grain kg^{-1} N. The AREn values for rice and wheat may vary from 30–39%.
3. Urea (*now* neem-coated urea) meets about 80% N of the total N needs of the country.
4. Part of basal N (N applied at planting) is deep placed in soil, while the rest is broadcast applied.
5. All top-dressed N is generally broadcast applied.
6. Nitrogen has positive interaction with most other plant nutrients. This is why response to N is higher with P or/and K application.
7. Fertilizer N may contribute up to ~ 6.36% of the total anthropogenic contribution of nitrous oxide.
8. Fertilizer N is responsible for nitrate pollution of shallow wells, and in some areas of India these waters already have nitrate levels above the USEPA limit of 10 mg NO_3–N $L^{-1.}$

REFERENCES

Bandyopadhyay, K.K. and Sarkar, M.C. 2005. Nitrogen use efficiency, ^{15}N balance and nitrogen losses in flooded rice in an Inceptisol. *Communications in Soil Science and Plant Analysis* **36**: 1,661–1,679.

Bhardwaj, A.S., Garg, S., Sondhi, S.K. and Taneja, D.S. 2012. Nitrate contamination of shallow aquifer groundwater in the central districts of Punjab. *Journal of Environment Science and Engineering* **54**(1): 90–97.

Bhuiyan, M.H.M., Ferdousi, M. and Iqbal, M.T. 2012. Foliar spray of nitrogen on raised bed increases yield of *aman* rice over conventional method. *ISRN Agronomy* **2012**. Article ID 184953 (doi:10.5402/2012/184953).

Biswas, P.P. and Sharma, P.D. 2006. A new approach for establishing fertilizer response ratio-the Indian Scenario. *Indian Journal of Fertilisers* **4**(1): 59–62.

Brennan, R.F. and Bolland, M.D.A. 2008. Significant nitrogen by sulphur interaction occurred for canola grain production and oil concentration in grain on sandy soils in the Mediterranean-type climate of southwestern Australia. *Journal of Plant Nutrition* **31**(7): 1,174–1,186.

Bruulsema, T.W., Fixen, P.E. and Sulkewski, G.D. (Ed.). 2012. *4R–Plant Nutrient Manual–A Manual for Improving Management of Plant Nutrition*. International Plant Nutrition Institute, Norcross, GA, USA.

Chen, Z., Wang, H., Liu, Z., Lu, D. and Zhou, J. 2016. The fate of ^{15}N labelled fertilizer in wheat–soil–system as influenced by fertilizer practices by a loamy soil. *Scientific Reports* **6**.3 4754 (doi: 10.1038/srep34754 (2016).

Chowdhary, N. 1989. *Consultancy Report on Environmental Pollution*. Department of Forests, Dhaka, Bangladesh.

Dobermann, A.R. 2005. Nitrogen use efficiency-state of the art. *Agronomy and Horticulture*, paper 316, University of Nebraska, Lincoln, USA (http://digitalcommons.unl.edu/agronomyfacpub/316).

Garg, A., Shukla, P.R., Ghosh, D., Kapshe, M. and Nair, R. 2003. Future GHG and local emission for India: Policy links and disjoints. *Mitigating and Adaptation Strategies for Global Changes* **8**: 71–92.

George, M. and Prasad, R. 1989. Studies on the effect of fertilizer N utilization by rice using ^{15}N technique in rice based multiple cropping systems. *Research and Development in Agriculture* **6**: 115–118.

Jat, M.L., Satyanarayana, T., Majumdar, K., Parihar, C.M., Jat, S.L., Tetarwal, J.P., Jat, R.K. and Saharawat, Y.S. 2013. Fertiliser best management practices for maize systems. *Indian Journal of Fertilisers* **91**(4): 80–94.

Knolbeloch, L., Saha, B., Hogan, A., Pestle, J. and Anderson, H. 2000. Blue babies and nitrate contaminated well water. *Environment and Health Perspectives* **108**(7): 675–678.

Kroetz, C. 1994. Nitrous oxide and global warming. *The Science of the Total Environment* **143**(2): 193–209.

Ladha, J.K., Pathak, H., Krupnik, T.J., Six, J. and van Kassel, C. 2005. Efficiency of fertilizer nitrogen in cereal production: Retrospect and perspectives. *Advances in Agronomy* **87**: 85–156.

Le, C., Zha, Y., Li, Y., Sun, D., Lu, H. and Yin, B. 2010. Eutrophication of lake waters in China: cost, causes and control. *Environment Management* **45**(4): 662–668.

Paul, T., Bisht, P.S., Pandey, P.C., Singh, D.K. and Roy, S. 2013. Rice productivity and soil fertility as influenced by nutrient management in rice (*Oryza sativa*)–wheat (*Tiricum aestivum*) cropping system. *Indian Journal of Agronomy* **58**(4): 495–499.

Prakasa Rao, E.V.S., Shivay, Y.S., Patra, D.D., Blaise, D. and Prasad, R. 2016. Indigenous materials for increasing efficiency of fertilizer nitrogen. *Indian Journal of Agronomy* **61** (4th International Agronomy Congress issue): 1–10.

Prasad, R and Shivay, Y.S. 2015. Fertilizer nitrogen for the life, agriculture and the environment. *Indian Journal of Fertilisers* **11**(8): 47–53.

Prasad, R and Shivay, Y.S. 2016. Deep placement and foliar fertilization of nitrogen for increased use efficiency- An overview. *Indian Journal of Agronomy* **61**(4): 71–75.

Prasad, R. 2009. Efficient fertilizer use: The key to food security and better environment. *Journal of Tropical Agriculture* **47**(1–2): 1–17.

Prasad, R. and Power, J.F. 1995. Nitrification inhibitors for agriculture, health and the environment. *Advances in Agronomy* **54**: 232–281.

Prasad, R., Singh, R.K., Rani, A. and Singh, D.K. 2000. Partial factor productivity of nitrogen and its efficiency in wheat. *Fertiliser News* **45**(5): 63–65.

Rajeswara Rao, B.R. and Prasad, R. 1987. Influence of competition irrigation levels and nitrogen fertilization on protein content and protein yield of three spring wheat cultivars. *Plant Foods for Human Nutrition* **37**:127–131.

Ravishankara, A.R., Daniels, J.S. and Portman, R.W. 2009. Nitrous oxide (N_2O): The dominant ozone-depleting substance emitted in the 21st Century. *Science* **326**(5949): 123–125.

Sharma, N.K. and Batra, P.K. 2011. A study on the fertilizer response ratios for various crops and cropping sequences. Project Report. Indian Agricultural Statistics Research Institute, New Delhi.

Shivay, Y.S., Pooniya, V., Prasad, R., Pal, M. and Bansal, R. 2016b. Sulfur-coated urea as a source of sulfur and enhanced efficiency of nitrogen fertiliser for spring wheat. *Cereal Research Communications* **44**(3): 513–523.

Shivay, Y.S., Prasad, R., Pooniya, V., Pal, M. and Bansal, R. 2016a. Effect of sulphur fertilization on biofortification of wheat (*Triticum aestivum*) grains with Fe, Zn, Mn and Cu. *Indian Journal of Agricultural Sciences* **86**(6): 823–824.

Smith, C.J., Freney, J.R., Sherlock, R.R. and Galbally, L.E. 1991. The role of urea nitrogen applied in a foliar spray to wheat at heading. *Fertilizer Research* **28**: 129–138.

Smith, K. (Ed.). 2010. *Nitrous Oxide and Climate Change*, Earthscan Ltd, London, UK.

Turner, R.E., Rabalais, N.N. and Justice, D. 2008. Gulf of Mexico–Alternate states and a legacy. *Environment Science and Technology* **42**(7): 2322327.

Varinder-pal Singh, Bijay-Singh, Yadvinder-Singh., Thind, H.S. and Gupta, R.K. 2010. Need based nitrogen management using the chlorophyll meter and leaf color chart in South Asia: A review. *Nutrient Cycling in Agroecosytems* **88**(3): 361–380.

Vitousek, P.M., Naylor, R., Crews, T., *et al.* 2009. Nutrient imbalances in agricultural development. *Science* **324**: 1,519–1,520.

Xiao-Hui, F., Yong-Sheng, S., De-Xi, L. *et al.* 2006. Ammonia volatilization losses and ^{15}N balance from urea applied to rice on a paddy soil. *Journal of Environmental Science* **18**(2): 299–303.

Zhou, S., Gu, B., Schesinger, W.H. and Ju, X. 2016. Significant accumulation of nitrate in Chinese semi-humid croplands. *Scientific Reports* **6**. 25088(2016)(doi: 10.1038/sep 25088).

QUESTIONS

1. Write down the equations for calculating the following

(a) AEn
(b) AREn
(c) PEn
(d) PFPn

2. Fill up the blanks

(a) DAP (diammonium phosphate) .... % N and .... % P_2O_5.
(b) CAN contains .... % N and .... % P_2O_5.
(c) Safe limit of NO_3–N in drinking water is .... mg L^{-1}.
(d) Using a knap-sack sprayer, generally available with farmers only .... kg N ha^{-1} can be applied in one spray.
(e) The second dose of N in wheat is applied at ....

3. Answer True (T) or False (F)

(a) Biological denitrification is the only process by which nitrous oxide is produced.
(b) In India, N application rates are very low and there is no danger of nitrate pollution of shallow wells in any part of the country.
(c) Application of P increases the crop response to N.
(d) Need-based N management is based on measuring the nitrate content in foliage.
(e) Crops generally take up about 75% of applied fertilizer N.

Chapter 8

Soil and Fertilizer Phosphorus

As compared to total N soil total P is present in much smaller amounts in soil; it is roughly about 10% of total N. Further, while N in soil is in the largest amount in surface 0 to 15 cm layer, soil P may be the same or in larger amounts in the sub-soil. Again, organic P constitutes only a small portion of total P. Thus, the Chemistry of soil P is generally centered on inorganic P. Total and organic P in some Indian soils is given in Table 8.1; hill soils had the highest and the desert soils the least total as well as organic P.

FRACTIONS OF INORGANIC P

All P in soils has originated from apatites [3 (Ca_3 $(PO_4)_2$.$Ca{\times}_2$, where $\times_2$ could be F, Cl, (OH) or CO_3). Large deposits of apatites form phosphate rocks, which are the major materials for making phosphate fertilizers. As the soil weathering proceeds, P in apatite reacts with Al, Fe and other cations and forms compounds of different solubility. A knowledge of these can help in understanding the P-supplying capacity of soils. A number of chemical extractants and resins are used for determining different fractions. In general, 7 fractions of inorganic soil P are mentioned. These are: saloid-P (soluble and loosely bound P extractable by NH_4Cl), labile P (determined by using anion-exchange resins), Al-P (NH_4F extractable), Fe-P (NaOH extractable), Ca-P (H_2SO_4 or HCl extractable), reductant-soluble- P (CDB or sodium citrate-sodium dithionite-sodium bicarbonate extractable) and occluded P (CTB followed by NaOH extractable) (Peterson and Corey, 1966; Zhang and Kovar, 2000; Tiessen and Moir, 2007). As would be expected saloid-P are the least of all the fractions. Also on calcareous and above neutral soils, Ca-P is most likely to be the highest. For example, Bhavsar *et al.* (2018) reported the following values (mg kg^{-1}) for the surface 0–15 cm Black soil of *p*H 7.5 from the Agricultural Botany Research Farm, Nagpur: total-P 485.0, inorganic-P 323.2, organic-P 161.8, Ca-P 195.9, Al-P 49.6, Fe-P 19.6, reductant soluble-P 33.7, occluded-P 19.4 and saloid-P 3.1.

Table 8.1. Total and organic phosphorus in some

Soil	Total P (mg kg^{-1})	Organic P (mg kg^{-1})	Organic P as % of total P
Hill	1,268	361.4	28.5
Black cotton	419	69.5	16.6
Alluvial	415	36.5	8.8
Red	376	87.6	23.3
Desert	240	15.6	6.5

Source: Dhir (1956); Kanwar and Grewal (1990)

Ionic forms of inorganic P (Pi) in soil

Ionic forms of inorganic P is controlled by soil *p*H and can be summed as below.

Ionic form	H_3PO →	H_2PO_4 →	HPO_4 →	PO_4^{-3}
Maximum between *p*H	0–1.0	4.0–6.0	8.0–10.0	14.0

In general, two ionic forms of P will dominate the soil solution at a given *p*H. For example at *p*H 3.0 it will be a mixture of H_3PO_4 and $H_2PO_4^{-1}$, while at *p*H 7.0 it will be a mixture of $H_2PO_4^{-1}$ and HPO_4. At *p*H above 11.0 it will be a mixture of HPO_4 and PO_4. From the nutrient availability point of view *p*H 6.0 to 6.5 is ideal.

The concentration of inorganic phosphorus (Pi) in soil solution is generally very low (~ 10 μM). As a contrast Pi concentration in root cells and xylem is hundreds to thousands fold (5–10 μM) (Lee and Radcliff, 1993). This steep concentration gradient helps plants in absorbing P from soil solution by diffusion, which itself is very slow (10^{-15}–10^{-10} m^2 s^{-1}) (Schachtman *et al.*, 1998). Thus, P accession from native soils by plants is by diffusion and is a slow process. As a contrast nitrates are highly water soluble and N is absorbed rapidly by mass flow.

FRACTIONS OF ORGANIC PHOSPHORUS

Only three fractions of organic P are identified, namely inositol hexaphosphate (phytic acid), nucleic acids (DNA, RNA) and phospholipids. Harrison (1987) after surveying the world literature, reported the following values: inositol hexaphosphate 1.4–356 mg kg^{-1} soil (0.3–62% of organic P), nucleic acids 0.1–97 mg kg^{-1} soil (0.1–65% of organic P), phospholipids 0.4–17 mg kg^{-1} soil (0.03–5.4% of organic P). In general, these three forms will account for 20–70% of organic P.

PHOSPHORUS FIXATION

The main reason for low apparent recovery efficiency of applied P (AREp) is that water-soluble P reacts rapidly with clay minerals and organic matter and either gets adsorbed on them when present in low concentrations or results in insoluble precipitates, when present in high concentrations, such as, 1 M ammonium/potassium/sodium phosphate (Prasad and Power, 1997).

Phosphorus fixation by clay minerals

The edges of clay minerals, especially the kaolinite group, have OH^- radicals on their edges and phosphate gets adsorbed on these radicals. When only one OH^- radical of H_3PO_4 is involved, the adsorption is said to be reversible, and when two OH^- radicals of H_3PO_4 are involved, the reaction is said to be irreversible (Hingston *et al.*, 1972). Concentrated solutions of phosphate can dissolve Al in the clay minerals and result in precipitates of Al-compounds.

Phosphorus fixation by hydroxides of aluminium and iron

Fixation of P on hydroxides of aluminium (Al) and iron (Fe) in tropical soils is similar to that in clay minerals. The only difference is that there is abundance of OH^- radicals and hence higher P- fixation. This would explain, why red and red and yellow soils have higher P-fixation capacity. Nad *et al.* (1975) reported P-fixing capacities of 72.6% in red and yellow soils, 61.6% in red soils, 59.2% in black soils and 46.5% in red and black soils of India.

Phosphorus fixation by calcium carbonate

Free calcium carbonate particles having some negative charge can adsorb P from dilute solutions, although it is less strongly bound than to Al and Fe hydroxides. With strong P solutions, however, insoluble calcium phosphates are formed and get precipitated.

Phosphorus fixation by organic matter

Soil organic matter is involved in fixation as well as release of P. Some researchers working with dilute P solutions have shown that organic matter complexes Ca, Fe and Al and this results in increased P availability.

Phosphorus reaction products

A number of researchers have tried to identify the insoluble reaction products formed in soil on the addition of phosphate fertilizers, mostly on the basis of solubility product principle (Wright and Peach, 1960). A number of mineral compounds have been identified, which include: Variscite ($AlPO_4.2H_2O$), Strengite ($FePO_4.2H_2O$), Brushite ($CaHPO_4.2H_2O$), Ammonium taranakite

($Al_2NH_4(PO_4)_2OH.2H_2O$), K-taranakite ($Al_5K_3H_6(PO_4)_8.18H_2O$), Octacalcium phosphate (Ca_8H_2 $(PO_4)_6.5H_2O$) and hydroxyl apatite ($Ca_{10}(PO_4)$ $(OH)_2$ (Sample *et al.*, 1986).

Factors affecting phosphorus fixation in soils

A number of factors affect P fixation in soils. These include *p*H, anions and cations present in soil and temperature (Prasad and Power, 1997).

Soil Test Methods for Available Phosphorus

There are quite a few methods for determining available P in soil. These include Bray's I (0.03 M ammonium fluoride + 0.25 M HCl) and Mehlich III (0.015 M ammonium fluoride + 0.2 M acetic acid + 0.25 ammonium nitrate + 0.013 M nitric acid) for acid soils and Olsen's method (0.5 M sodium bicarbonate) for calcareous and alkaline soils (Fixen and Grove, 1990). In India, Olsen's method (Olsen *et al.*, 1954) is used for determining available P and has correlated well with response of crops to P (Tandon, 1987). The limits for low, medium and high available Olsen's P used by soil testing laboratories in India are: below 10, 10–24.6 and above 24.6 kg ha^{-1} respectively (Subba Rao, 1995). The available P status of Indian soils is shown in Fig. 8.1.

APPARENT RECOVERY EFFICIENCY OF PHOSPHORUS

Apparent recovery efficiency of P (AREp) (for definition reference may be made to section Nitrogen-use efficiency) in crops can vary from 11 to 29%, depending on crop, rate of application, soil and growing conditions. Some data on AREp are given in Table 8.2.

Table 8.2: Apparent recovery efficiency of P in some crops

Location (Soil type)	Available P (kg ha^{-1})	Crop	Fertilizer P (kg P ha^{-1})	AREp (%)	Reference
Delhi (Sandy loam)	12.8	Cotton	17.5	16	Sepat *et al.* (2012)
			35	18	
	12.8	Groundnut	17.5	12	Sepat *et al.* (2012)
			35	11	
	10	Wheat	17.2	17	Singh and Ahlawat (2007)
			34.4	16	
Varanasi (Sandy clay loam)	18.5	Pigeonpea	25	18	Singh and Singh (2012)
			50	21	
			75	16	
Udaipur (Clay loam)	20.9	Maize	20	28	Mehta *et al.* (2005)
			40	14	
			60	12	

Source: Blaise *et al.* (2014)

PHOSPHORUS MANAGEMENT

The raw material required for the manufacture of P fertilizers is non-renewable phosphate rock (PR), most of which has to be imported in India. Five major countries with PR reserves [Million metric tonnes (MMT) of P_2O_5] are: Morocco (51,000), China (3,700), the USA (1,800), Jordan (900) and Russia (500) (Van Kauwenbergh, 2010) and the total global reserve is estimated at 60,000.

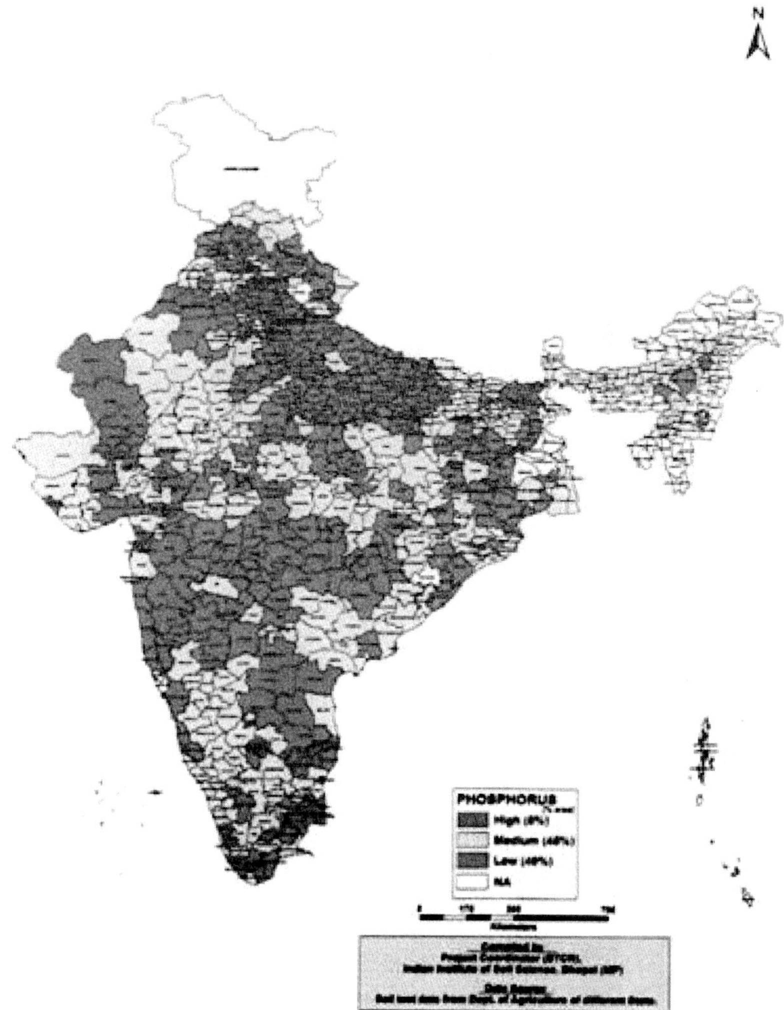

Fig. 8.1: Available P status of Indian soils. (with permission from Dr Pradip Dey, Project Coordinator, STCR, Indian Institute of Soil Science, Bhopal)

MMT of P_2O_5. Continued availability of phosphate rock (PR) over time has been a matter of global concern, since the world's total phosphorus reserves may last only for about 380 years at the current rate of production and use of P fertilizers; in reality this could be much less, about 300 years, because the increased demand for food production would require an increased production of P fertilizers. Supply of P is likely to peak around 2030 and decline to one-third by 2100 (Cordell *et al.*, 2009). Efficient phosphorus management in crop production is therefore important (Prasad *et al.*, 2016). The 4Rs of fertilizer management as discussed for N management are equally important for P.

Sources of P

There are only three straight P fertilizers, namely, single or ordinary superphosphate (SSP) (16% P_2O_5), triple or concentrated superphosphate (TSP) (46% P_2O_5), and low-grade Mussoorie rock phosphate (18–25% P_2O_5). There are a number of NP and NPK fertilizers (Table 7.1). Of the three straight P fertilizers, only SSP is produced in large enough amounts. It is losing ground against a tough competition with diammonium phosphate (DAP), which meets about two-thirds of the country's P needs. The main reason is SSP's low P content of 16% P_2O_5 as against 46% P_2O_5 in DAP. So for applying P at the rate of 23 kg P_2O_5 ha^{-1} to a field, a farmer has to carry 3 bags of 50 kg fertilizer (all fertilizer except urea in India is sold in 50 kg bags) of SSP as against only 1 bag of DAP. This brings a considerable saving in transportation cost. Also SSP used to be produced in powder form (now quite a few producers are granulating it) and it chocked the chutes of fertilizer drills. Further, because of some free sulphuric acid in it, the farmers' applying it through broadcast method had skin irritation in hands. Nevertheless, it has 11% S and when a farmer applies 23 kg P_2O_5 ha^{-1}, he also applies 16.5 kg S ha^{-1}, which is quite near to the amount of S recommended for pulses and oilseeds. When used over a long period, SSP could also have advantage over DAP owing to the presence of calcium sulphate in it. The DAP on the other hand has the advantage of meeting the entire N and P need of pulses, when two bags of it are applied per hectare (supplying 18 kg N and 46 kg P_2O_5).

Manufacture of nitrophosphates (NPHs) was started in Europe to avoid the use of sulphuric acid, since there are no large deposits of S in Europe. In the manufacture of NPHs, phosphate rock is reacted with nitric acid in place of sulphuric acid. Since one of the products in the manufacture of NPHs is calcium nitrate, which is highly hygroscopic, it has to be removed. In ODDA process, it is removed by cooling, crystalizing and centrifuging. Calcium nitrate is then marketed as a nitrogen fertilizer 'Nitrochalk' and is good for acid soils. In PEC process, calcium nitrate is converted to calcium chloride by reacting it with HCl. Some of the Ca in calcium chloride reacts with mono-calcium phosphate and leads to the formation of dicalcium phosphate, which is not water soluble (WSP), but is ammonium citrate soluble (ACSP). Thus, PEC nitophosphate contains both WSP and ACSP (discussed later in this section).

As regards low-grade Mussoorie phosphate rock, research has shown that it can be directly used in acid soils, especially in rice (Sidaramappa *et al.*, 1991) or after partial acidulation (Panda and Misra, 1970). On neutral and above-neutral soils it can be used as a mixture with triple super phosphate (Govil and Prasad, 1972) or pyrite (Sharma and Prasad, 1996), as phosphocompost (Singh, 2012) or with phosphate-solubilizing bacteria (PSB) (Sharma and Prasad, 2003; Kumari and Phogat, 2008). However, Mussoorie rockphosphate is no longer in the market.

Efforts are also underway to develop slow-release polymer-coated fertilizers (Sanders *et al.*, 2012) and nano fertilizers (Subramanian *et al.*, 2015) for increasing AREp.

Water solubility of P in P-fertilizers: Phosphorus is present in fertilizers in three forms: *1*, Water-soluble P (WSP) [mono calcium phosphate Ca $(H_2PO_4)_2$, mono ammonium phosphate $(NH_4)\ H_2PO_4$ diammonium phosphate $(NH_4)_2HPO_4$), mono potassium phosphate (KH_2PO_4)]; *2*, Citrate-soluble [dicalcium phosphate $(CaHPO_4)$]; *3*, Insoluble P [apatites or ground phosphate rock and tricalcium phosphate $(Ca_3\ (PO_4)_2$]. In the analysis, a sample of P fertilizer is first extracted with water (WSP), the residue left is then extracted with 1 M ammonium citrate (ACSP). The residue finally left is considered as insoluble P. The sum of WSP and ACSP is known as available P, since plants take up P both as $H_2PO_4^{-1}$ and HPO_4^{-2}. Some agronomic experiments were conducted in India for comparing ODDA nitrophosphate (containing all WSP) and PEC nitrophosphate (containing 30% WSP). The results showed that for neutral and above-neutral *p*H soils, ODDA nitrophosphate was superior, while on acid soils PEC nitrophosphate was equally good or even better (Sinha and Dutta, 1970; Prasad *et al.*, 1971; Mahapatra *et al.*, 1973, Prasad and Dixit, 1976).

Rate of application

Most P is applied on the basis of soil test and 0.5 M sodium bicarbonate extracted P has served very well.

Timing and method of application

Most P is applied at sowing and is placed using a fertilizer drill or is broadcast and incorporated at final land preparation. Foliar application of P could be of advantage. Recent studies on greengram at Allahabad (Dixit and Elamathi, 2007), on soybean [*Glycine max* (L.) Merill] at Coimbatore (Vinod Kumar *et al*., 2013) and in frontline demonstrations in Bihar, Jharkhand and Bengal on pigeonpea [*Cajanus Cajan* (L.) Millsp.] (ATARI, 2017) have proven the advantage of 2% DAP spray over and above the recommended basal P application.

RESIDUAL EFFECT OF P ON AVAILABLE PHOSPHORUS AND YIELDS OF SUCCEEDING CROPS

Since, in general, about 30–60 kg P_2O_5 ha^{-1} is applied to a crop in India; one should not expect high residual effects. In rice–wheat cropping system, which covers about 10 million ha in north India, Gill and Meelu (1983) from Ludhiana, Punjab, suggested that if 26 kg P ha^{-1} (60 kg P_2O_5 ha^{-1}) is applied to wheat, rice could be grown without application of P. However, from the same research centre, Yadvinder–Singh *et al.* (2000) recommended the application of 32 kg P ha^{-1} to wheat and 15 kg P ha^{-1} to rice in the RWCS. Kumar and Yadav (2001) on the basis of a long-term experiment at Faizabad, Uttar Pradesh, India, on RWCS also reported that, both rice and wheat responded well to P application. Prasad (2007) also observed that the residual effect of P applied to wheat was not enough and both rice and wheat need to be fertilized with P. The effects of 25 years of annual applications of P fertilizer on the accumulation and migration of soil Olsen-P, and the effects of soil residual P on crop yields by withholding P application for the following 5 years, were evaluated on a sandy-loam soil at Ludhiana. Annual application of P fertilizer for 25 years to crops in summer (groundnut), winter (wheat, mustard or rapeseed) or in both seasons raised the Olsen-P status of the plough layer (0–15 cm) from initially very low (12 kg P ha^{-1}) to medium (18 kg P ha^{-1}) and very high levels (40–59 kg P ha^{-1}) depending on crop uptake and rates of P application (Aulakh *et al*., 2007). In the following 5 years, the raising of 10 crops without P fertilizer applications decreased the accumulated Olsen-P by only 20–30% depending on the amount of accumulated P and crop requirements. These results demonstrated little agronomic advantage above approximately 20 mg kg^{-1} Olsen-P build up.

INTERACTION WITH OTHER NUTRIENTS

Positive or synergistic reaction of P with N and K has already been discussed and it forms the basis of balanced NPK fertilization. As a matter of facts, all essential nutrients when present in needed amounts, should interact positively (synergistic effect), because that would be the basis of essentiality. However, when an essential element is present in excess than needed, it creates problems. Phosphorus, due to its reactivity and tendency to form insoluble compounds with most metallic cations, can lead to negative interaction (antagonistic effect). One of the most often discussed is the P ×

Zn negative interaction. For example, Soltangheisi *et al.* (2014) reported that, as the P application was increased the Zn concentration in sweet corn leaves decreased (Table 8.3). It is reported that the presence of high amounts of P in soil solution inhibits translocation of Zn from roots to leaves (Nair and Babu, 1975).

Table 8.3: Zn concentration in sweet corn leaves as influenced by P application rate

Zn + P application (mg kg^{-1}soil)	Zn concentration in sweet corn leaves (μg g^{-1})
0 + 0	108.7
0 + 50	59.7
0 + 100	56.7
0 + 200	54.3
5 + 0	239.3
5 + 50	180.7
5 + 100	142.3
5 + 200	118.3

Source: Soltangheisi *et al.* (2014)

However, a judicious application of P and Zn can change this negative interaction into a positive interaction as shown by data on corn (maize) from Kansas in the USA (Table 8.4) (Anonymous, 1999).

PHOSPHATE FERTILIZERS AND P EUTROPHICATION OF WATERS

Since P reacts fast with the cations in soil solution and also gets fixed on the surfaces of clay minerals and hydroxides of Al and Fe, one may not expect leaching loss of applied P, especially if the doses are low as in India. However, globally there has been a concern about P eutrophication of lake and marine waters (Chien *et al.* 2011). Phosphorus is perhaps more critical than nitrogen for some species of algae, such as blue green algae (*Cyanobacteria*). Algal blooms lead to the depletion of dissolved oxygen in water (hypoxia), which can lead to fish mortality, and their death and decomposition impair such waters for drinking, recreation, and industry (Foy, 2005). Phosphorus from soils is lost as erosion and surface runoff. As would be expected, higher rates of P application result in higher loss of P. In China in rice, an application of 0, 25, 60, 120, and 240 kg P ha^{-1} resulted in a loss of 0.13, 0.50, 0.94, 3.02, and 5.97 kg P ha^{-1} through runoff water (Xia *et al.*, 2008). The period of most risk of P loss from rice field was within 1–2 months after P application (Zhang *et al.*, 2007). Kimmel *et al.* (2001) reported that broadcast P application increased P losses in a grain sorghum [*Sorghum bicolor* (L.) Moench]–soybean system in the USA. It is also reported that P fertilizers having no WSP, such as ground phosphate rock, can minimize the runoff losses of P as compared to WSP fertilizers (Shigaki *et al.*, 2006).

Table 8.4: Optimizing yield of corn by judicious application of phosphorus and zinc (data from Kansas, USA)

Zn + P application ($kg\ ha^{-1}$)	Corn yield ($Mg\ ha^{-1}$)
0 + 0	8.2
0 + 89.6	7.5
22.4 + 0	6.8
22.4 + 89.6	11.0

Source: Anonymous (1999)

SUMMARY

1. As a contrast to N bulk of P occurs in inorganic form.
2. Total P in Indian soils could vary from 240 to 1,268 mg kg^{-1} soil, being the highest in Hill soils and the lowest in Desert soils.
3. Organic P expressed as the percentage of total P could vary from 6.5 to 28.5, being the highest in Hill soils and the lowest in Desert soils.
4. The fractions of inorganic P are: saloid-P, labile P, Al-P, Fe-P, Ca-P, reductant-soluble-P and occluded P.
5. The major fractions of organic P are: inositol hexaphosphate, nucleic acids and phospholipids.
6. Applied P in soils can be fixed by silicate clays, hydroxides of Al and Fe, calcium carbonate and organic matter.
7. A number of mineral compounds have been identified as reaction products of the fertilizer P after its application in soil. These include: Variscite, Strengite, Brushite, Ammonium taranakite, K-taranakite, Octacalcium phosphate and Hydroxyl apatite.
8. Factors affecting P fixation in soils are *p*H, anions and cations present in soil and temperature.
9. Phosphorus fertilizers in India are: SSP, TSP, DAP, UAP, APS, nitrophosphates, NPKs.
10. Besides considerable research Mussoorie rock phosphate is no longer used as a P fertilizer in India.
11. All P fertilizers in India except some nitrophosphate have all their P as water soluble.
12. Phosphorus in most crops is applied at sowing.
13. In most crops except rice, P is deep placed in soil or it is broadcast applied and incorporated in soil at sowing/ transplanting.
14. There have been some successful attempts of foliar application of DAP on pulse crops.
15. Not much residual effect of P has been reported in India.

16. In general P and Zn interact negatively; however, a judicious application can make this interaction positive.
17. Surface application of high rates of P can lead to erosion and surface-runoff losses of P.

REFERENCES

Anonymous. 1999. Phosphorus interactions with other nutrients. *Better Crops* **83**(1): 11–13.

ATARI. 2017. Cluster frontline demonstrations on pulses in Bihar, Jharkhand and West Bengal, ICAR–Agricultural Technology Application Research Institute, Kolkata (email: atarikalkata@gmail.com)

Aulakh, M.S., Garg, A.K. and Kabba, B.S. 2007. Phosphorus accumulation, leaching and residual effects on crop yields from long term applications in the subtropics. *Soil Use and Management* **23**(4): 417–427.

Bhavsar, M.S., Ghagare, R.B. and Shinde, S.N. 2018. Study of different phosphorus fractions and their relationship with soil properties in Agricultural Botany Research Farm, Nagpur, India. *International Journal of Current Microbiology and Applied Sciences* **7**(1): 1,130–1,137.

Blaise, D., Venugopalan, M.V. and Singh, G. 2014. Phosphorus management. (In) *Textbook of Plant Nutrient Management,* pp. 93–121. Prasad, R., Kumar, D., Rana, D.S., Shivay, Y.S. and Tewatia R.K. (Eds). Indian Society of Agronomy, New Delhi.

Chien, S.H., Prochnow, L.I., Tu, S. and Synder, C.S. 2011. Agronomic and environmental aspects of phosphate fertilizers varying in source and solubility: An update review. *Nutrient Cycling in Agroecosystems* **89**: 229–255.

Cordell, D., Drangert, J.O. and White, S. 2009. The story of phosphorus: global food security and food for thought. *Global Environment Change* **19**: 292–305.

Dhir, R.P. 1956. Phosphorus status of major Indian soils. Thesis Associate, Indian Agricultural Research Institute, New Delhi.

Dixit, P.M. and Elamathi, S. 2007. Effect of foliar application of DAP, micronutrients and NAA on growth and yield of green gram (*Vigna radiata* L.). *Legume Research* **30**(4): 305–307.

Fixen, P.E. and Grove, J.H. 1990. Testing soils for phosphorus. (In) *Soil Testing and Plant Analysis*, edn 3, pp: 141–180. Westerman, R.L., (Ed.), Soil Science Society of America, Madison, USA, Book Ser.

Foy, R.H. 2005. The return of phosphorus paradigm: Agricultural phosphorus and eutrophication. (In) *Phosphorus in Agriculture and the Environment* pp: 911–939. Agronomy Monograph No. 46. Sims, J.T. and Sharpley, A.N. (Eds). American Society of Agronomy, Madison, WI, USA.

Gill, H.S. and Meelu, O.P. 1983. Studies on the utilization of phosphorus and causes for its differential response in rice–wheat rotation. *Plant and Soil* **74**: 211–222.

Govil, B.P. and Prasad, R. 1972. Growth characters and yield of sorghum as affected by contents of water soluble P in triple superphosphate/ dicalcium phosphate and triple superphosphate/ rock phosphate mixtures. *Journal of Agricultural Science,* Cambridge **79**: 485–492.

Harrison, A.F. 1987. *Soil Organic Phosphorus–A Review of World Literature*, CAB International, Wallingford, UK.

Hingston, F.J., Posner, A.M. and Quirk, J.P. 1972. Anion adsorption by goethite and gibbsite. I. The role of protons in determining adsorption envelops. *Journal of Soil Science* **23**: 177–192.

Kanwar, J.S. and Grewal, J.S. 1990. Phosphorus fixation in Indian soils – a review. Indian Council of Agricultural Research, New Delhi.

Kimmell, R.J., Pierzynski, G.M., Janssen, K.A. and Barn, P.L. 2001. Effects of tillage and phosphorus placement on phosphorus runoff losses in a grain sorghum–soybean rotation. *Journal of Environmental Quality* **30**: 1,324–1,330.

Kumar, A., and Yadav, R.L. 2001. Long-term effects of fertilizers on soil fertility and productivity of a rice–wheat system. *Journal of Agronomy and Crop Science* **186**: 47–54.

Kumari, K. and Phogat, V.K. 2008. Rock phosphate, its availability and solubilization in the soil–A review. *Agricultural Research* **29**(2): 108–116.

Lee, R.B. and R.G. Radcliffe. 1993. Subcellular distribution of inorganic phosphate, and levels of nucleoside triphosphate, in mature maize roots at low external phosphate concentrations: Measurements with ^{31}P- NMR. *Journal of Experimental Botany* **44**: 587–598.

Mahapatra, I.C., Prasad, R. and Leelvathi, C.R. 1973. Efficiency of nitrophosphate as compared to straight N and P fertilizers. (In) Proceedings of the *ISMA/FAI Seminar on Nitrophosphate and other NPK Fertilizers,* Fertiliser Association of India, New Delhi. 1–2/1–19.

Mehta, Y.K., Shakavat, M.S. and Singh, S.M. 2005. Influence of sulphur, phosphorus and farmyard manure on yield attributes and yield of maize (*Zea mays*) in Southern Rajasthan conditions. *Indian Journal of Agronomy* **50**(3): 203–205.

Nad, B.K., Goswami, N.N. and Leelawathi, C.R. 1975. Some factors influencing the phosphorus fixing capacity of Indian soils. *Journal of the Indian Society of Soil Science* **32**: 288–291.

Nair, K.P.P. and Babu, R.G. 1975. Zinc–phosphorus interaction studies in maize. *Plant and Soil* **42**: 517–536.

Olsen, S.R., Cole, C.V., Watanabe, F.S. and Dean, L. 1954. Estimation of available phosphorus in soil by extraction with sodium carbonate. USDA Circular 933. USDA, Washington DC, USA.

Panda, N. and Misra, V.K. 1970. Use of partially acidulated rock phosphate as a possible means of minimizing phosphorus fixation in acid soils. *Plant and Soil* **33**: 226–234.

Peterson, C.W. and Corey, R.B. 1966. A modified Cheng and Jackson for routine fractionation of inorganic phosphates. *Soil Science Society of America Proceedings* **30**: 563–565.

Prasad, R. 2007. Phosphorus management in the rice–wheat cropping system of the Indo-Gangetic Plains. *Better Crops-India* **1**(1): 8–11.

Prasad, R. and Dixit, L.A. 1976. *Fertilizers containing partially water soluble or no water soluble phosphate*. Bulletin, Indian Council of Agricultural Research, New Delhi, 34 p.

Prasad, R. and Power, J.F. 1997. *Soil Fertility Management for Sustainable Agriculture*, 356 p. CRC Press, Boca Raton, FL, the USA.

Prasad, R. Bhendia, M.L. and Turkhede, B.B. 1971. Relative efficiency of phosphatic fertilizers in different soils of India. *Proceedings of International Symposium* on *Soil Fertility Evaluation*, New Delhi **1**: 747–756.

Prasad, R., Shivay, Y.S., Majumdar, K. and Prasad, S. 2016. Phosphorus management. (In) *Soil Phosphorus,* pp: 81–113. Lal, R. and Stewart, B.A. (Eds.), CRC Press, Boca Raton, FL, USA.

Sample, E.C., Soper, R.J. and Racz, G.J. 1986. Reactions of phosphate fertilizers in soils. (In) *The Role of Phosphorus in Agriculture*. 2nd Print, Khasawneh F.E., Sample, E.C. and Kamprath, E.J. (Eds.),American Society of Agronomy, Madison, WI, USA.

Sanders, J.L., Murphy, L.S., Noble, A., Melgar, R.J. and Perkins, J. 2012. Improving phosphorus use efficiency with polymer technology. *Procedia Engineering* **46**: 176–184.

Schachtman, D.P., Reid, R.J. and Ayling, S.M. 1998. Phosphorus uptake by plants: From soil to cell. *Plant Physiology* **116**(2): 447–453.

Sepat, S., Ahlawat, I.P.S. and Rana, D.S. 2012. Effect of phosphorus sources and levels on Bt cotton (*Gossypium hirsutum*)-based cropping systems. *Indian Journal of Agronomy* **57**(3): 235–240.

Sharma, S.N. and Prasad, R. 1996. Mussoorie rock phosphate pyrite mixture as phosphate fertilizer. *Fertilizer Research* **45**: 187–191.

Sharma, S.N. and Prasad, R. 2003. Yield and P uptake by rice and wheat grown in a sequence as influenced by phosphate fertilization with diammonium phosphate and mussoorie rock phosphate with or without crop residues and phosphate solubilizing bacteria. *Journal of Agricultural Science*, Cambridge **141**: 359–369.

Shigaki, F., Sharpley, A. and Pochnow, L.I. 2006. Source related transport of phosphorus in surface runoff. *Journal of Environmental Quality* **35**: 2,229–2,235.

Siddaramappa. R., Jagadish, N.R. and Srinivasamurthy, C.A. 1991. Efficiency of rock phosphate as phosphatic fertilizer to rice in acid soils of Karnataka. (In) *Plant–Soil Interactions at low pH,* Wright, R.J., Baligar, V.C. and Murrmann, R.P., (Eds). *Developments in Plant and Soil Sciences* **45**: 307–312, Springer, Dordrecht, Netherlands.

Singh, A.K. and Singh, R.S. 2012. Effect of phosphorus and bioinoculants on yield, nutrient uptake and economics of long-duration pigeonpea (*Cajanus cajan*). *Indian Journal of Agronomy* **57**(3): 265–269.

Singh, C.P. 2012. Preparation of phosphocompost and its effect on the yield of moongbean and wheat. *Biological Agriculture and Horticulture* **2**(3): 223–229.

Singh, U. and Ahlawat, I.P.S. 2007. Phosphorus management in pigeonpea (*Cajanus cajan*)–wheat (*Triticum aestivum*) cropping system. *Indian Journal of Agronomy* **52**(1): 21–26.

Sinha, A.K. and Dutta, N.P. 1970. Efficiency of PEC nitrophosphates in Bihar. *Technology* **7**(2): 42–44.

Soltangheisi, A., Rahman, Z.A., Ishak, C.Z., Musa, M.H. and Zakikhani, H. 2014. Interaction effects of phosphorus and zinc on their uptake and ^{32}P absorption and translocation in sweet corn (*Zea mays* var Sachharata) grown in a tropical soil. *Asian Journal of Plant Sciences* **13**(3): 129–135.

Subba Rao, A. 1995. Analysis of soils for available major nutrients. (In) *Methods of Analysis of Soils, Plants, Waters and Fertilizers,* pp: 13–35. Tandon, H.L.S. (Ed.), Fertilizer, Consultation and Development Organization (FDCO), New Delhi.

Subramanian, K.S., Manikandan, A., Thirunavukkarasu, M. and Rahale, C.S. 2015. Nano-fertilizers for balanced crop nutrition. (In) *Nanotechnologies in Food and Agriculture,* pp: 69–80. Rai M., Riebeiro C., Mattoso L and Duran N. (Eds), Springer.

Tandon, H.L.S. 1987. *Phosphorus Research and Agricultural Production in India,* edn 3, 140 pp. The Sulfur Institute, Washington, DC, USA.

Tiessen, H. and Moir, J.O. 2007. Characterization of available phosphorus by sequential extraction. (In) *Soil Sampling and Methods of Analysis* edn 2. pp: 203–306. Carter, M.R. and Gregorich, E.G. (Eds.), CRC Press, Boca Raton, FL, the USA.

Van Kauwenbergh, S.J. 2010. *World Phosphate Rock Reserves and Resources,* 60 pp. International Fertilizer Development Centre (IFDC), Muscle Shoals, AL, the USA.

Vinod Kumar, C., Vaiyapuri, K., Amanullah, M.A. and Gopalaswamy, G. 2013. Influence of foliar spray of nutrients on yield and economics of soybean [*Glycine max* (L.) Merill]. *Journal of Biological Sciences* **13**(6): 363–365.

Wright, B.C. and Peech, M. 1960. Characterization of phosphate reaction products in acid soils by the application of solubility criteria. *Soil Science* **90**: 32–43.

Xia, T.X., Li, W.C. and Pan, J.Z. 2008. Risk assessment on soil environment quality and losses of nitrogen and phosphorus for the gravel soils under different farming practices in the watershed of Lake Fuxian. *Journal of Lake Science* **20**: 110–116.

Yadvinder-Singh, Dobermann, A., Bijay-Singh., Bronson, K.F. and Khind, C.S. 2000. Optimal phosphorus management strategies for wheat–rice cropping on a loamy sand. *Soil Science Society of America Journal.* **64**: 1,413–1,422.

Zhang, H. and Kovar, L. 2000. (In) *Methods of Phosphorus Analysis for Soils, Sediments, Residuals, and Waters.* Pierzynski, G.M. (Ed.). North Carolina State University, Raleigh, NC, USA, Southern Cooperative Series Bulletin No. 396. (http://www.soil.ncsu.edu/sera17/publications/sera17-2/pm_cover.htm).

Zhang, Z.J., Zhang, J.Y., He, R. and Zhu, Y.M. 2007. Phosphorus interception in floodwater of paddy field during the rice growing season in Tai Lake Basin. *Environmental Pollution* **145**: 425–433.

QUESTIONS

1. Briefly discuss

(a) Water soluble P in P-fertilizers
(b) Reductant soluble P in soil
(c) Inositol hexaphosphate
(d) Nitrophospates
(e) Residual P in soil

2. Match the followings

(a)	Al-phosphate	(i)	Sodium bicabonate
(b)	Nucleic acids	(ii)	Nitric acid
(c)	Phosphate rock	(iii)	Variscite
(d)	Available P	(iv)	Organic P
(e)	Nitrophosphate	(v)	PSB

3. Answer True (T) or False (F)

(a) Fe is available in high amounts in soils < *p*H 5.0
(b) Fixation of P in red and lateritic soils is due to the presence of OH^- radicals.
(c) SSP is a good source of S.
(d) Nitrophosphates may have some citrate-soluble P.
(e) There is no need for slow release P fertilizers.

Chapter 9

Soil and Fertilizer Potassium

Out of the three primary plant nutrients (NPK), potassium (K) is taken up by crop plants in the largest amount, yet as compared to N and P, only a tiny fraction of it goes in the grain or seed. For example, Shivay *et al.* (2015) reported from a study on rice that about 20 kg N, 6 kg P (13.7 kg P_2O_5) and 46 kg K (55.2 kg K_2O) was taken up by the crop to produce each Mg or tonne (1,000 kg) of rice kernel; however, only 4.3% of K taken up by the crop was found in the rice kernel as compared to 44% of N and 66% of P taken up by the crop. Some information on N, P and K removed per tonne of grain/ seed for some crops is given in Table 9.1.

Table 9.1: Comparative removal of nitrogen, phosphorus (as P_2O_5) and potassium (as K_2O) per tonne of grain/seed in some crops (All values rounded up to whole numbers)

Crop	kg N	kg P_2O_5	kg K_2O	P_2O_5 and K_2O uptake in relation to N		
				N	P_2O_5	K_2O
Rice	20.0	11.0	30	100	45	150
Wheat	25.0	9.0	33	100	35	134
Chickpea	46.0	8.0	50	100	18	107
Pigeonpea	64.0	18.0	42	100	28	66
Groundnut	58.0	20.0	30	100	34	52
Soybean	67.0	18.0	44	100	26	66
Mustard	33.0	16.0	42	100	50	127
Sugarcane (cane)	1.7	0.2	2	100	12	117
Potato (tuber)	3.9	1.4	5	100	36	126

Source: Tiwari (2014)

Thus, based on the data in Table 9.1, removal of K_2O (kg ha^{-1}) for above crops will be: Rice (6 t ha^{-1}) 180; wheat (6 t ha^{-1}) 198; chickpea (2.5 t ha^{-1}) 125; pigeonpea (2.5 t ha^{-1}) 110; groundnut (10 t ha^{-1}) 300; soybean (2.5 t ha^{-1}) 110; mustard–*Brassica juncea* (L.) Czern. (2 t ha^{-1}) 84; sugarcane (80 t ha^{-1}) 160; potato–*Solanium tuberosum* L. (25 t ha^{-1}) 125.

SOIL POTASSIUM

Most potassium (K) in soil is present in inorganic form and in Indian soils it could vary from 200 mg kg^{-1} to 2.5 g kg^{-1} or even more. Potassium is present in soil in four major forms, namely mineral K, non-exchangeable K, exchangeable K and solution K, which tend to be in equilibrium.

Mineral potassium/ structural potassium/lattice potassium

The K-bearing primary minerals in soil are silicate minerals, namely orthoclase feldspar or K-feldspar [(K, Na)Al Si_3O_8] and microcline [(Na, K) Al SiO_4], generally present in the coarse sandy fraction. Micas, muscovite–the Al-mica [KAl_3 Si_3O_{10} $(OH)_2$], biotite – the Fe mica [K (Fe, Mg)$_3$ Si_3O_{10} $(OH)_2$] and phlogopite – the Mg mica [KMg_3 Si_3O_{10} $(OH)_2$] are present in the clay fraction. The names of micas are derived from the cation in the octahedral layer, which could be Al, Fe or Mg. In biotite there is some mixture of Fe and Mg. As is obvious, K is in the structure of the minerals and is released only when the structure is disturbed by weathering. Due to likely charge changes in iron from Fe^{3+} to Fe^{2+} the biotite mica is more fragile and liberates K faster than muscovite and phlogopite micas.

Non-exchangeable or fixed K

When the primary minerals weather into secondary minerals, such as illite or vermiculite, potassium (K) is released from the primary minerals and is held tightly between tetrahedral layers of the secondary minerals and is therefore also called as fixed K. Only ions such as NH_4^+ can replace fixed K, because both have nearly same ionic radii. Hydrated Ca^{2+} or Mg^{2+} cannot replace K, because of their large radii. Non-exchangeable K (NEK) forms a fairly large part of soil-K pool (Table 9.2) and is sparingly to mildly available. Some crops, such as rye grass (*Lolium perenne* L.) have root-system that can feed on NEK. The NEK is determined in soil by extracting it with boiling 1 N nitric acid.

Table. 9.2: Potassium in different fractions (% of total K) in some Indian soils

Soil	Water soluble (% of total K)	Exchangeable (% of total K)	Non-Exchangeable (% of total K)	Total K C (+) mole kg^{-1} (g kg^{-1})
Alluvial Calcareous illitic	0.12	0.28	7.52	57.43 (2.24)
Acidic kaolinitic	0.32	0.78	2.34	10.25 (0.40)
Vertisol–Verticsmectitic	0.25	2.12	5.56	35.38 (1.38)
Red and lateritic Kaolinitic	0.19	0.49	3.14	26.41 (1.03)

Source: Sekhon *et al*. (1992).

Exchangeable potassium

Exchangeable potassium (K) is held in the exchange complex of the clay minerals and is more in the 2 : 1 layer silicates, such as, smectites than 1 : 1 layer silicates, such as, kaolinite. This is because 2 : 1 layer silicates have higher CEC due to isomorphous substitution (See section 1.3.1).

Soil solution potassium

This K is present in soil solution at any time. It is the smallest fraction. Relative values for different K fractions in some Indian soils are given in Table 9.2.

In a recent study in sub-montane soils of Maharashtra, exchangeable-K and non-exchangeable-K contributed 2.87 and 8.97% of total K, respectively, the highest values were noticed in Vertisols followed by Inceptisols and Entisols. Total K in these soils ranged from 4.2 to 7.3 g kg^{-1} soil (Annapurna *et al.* 2017) (Fig. 9.1).

Out of the above four forms of soil K, three tend to be in equilibrium as shown in the expression below.

Mineral-K ⟶ Non-exchangeable-K (NEK) ⟷ Exchangeable-K (EK) ⟷ Soil solution-K

Fig. 9.1: Potassium dynamics in soil

When soil solution-K is depleted by crop removal or by leaching, part of exchangeable-K is released to soil solution. Similarly, as and when soil-exchange complex is depleted of K, some K is released from mineral K.

Potassium Fixation in Soils

Figure 9.1 shows that some of the processes are reversible. Thus, when K fertilizer is applied to soil, some of the K gets on the exchange complex of clay minerals and finally gets fixed. Fixation of K is higher in Vertisols, which are richer in 2 : 1 layer silicates. Illites, vermiculites and smectites fix more K than kaolinite. For example, Mitra *et al.* (1958) reported that in a laboratory study illite fixed 52.5% of applied K, while bentonite fixed 12.7% and kaolinite fixed only 7.6% of applied K. However, so fixed K becomes later available to crops. This is thus different than P-fixation of applied, which becomes unavailable to crops. As a matter of fact, fixation of K prevents its loss due to leaching, and is thus a good property of soil.

SOIL TEST FOR AVAILABLE K IN SOIL

Available K in soil is determined by extraction with 1 N ammonium acetate and the limits are as under (Muhr *et al.*, 1965):

Low – less than 120 kg K ha^{-1} (144 kg K_2O ha^{-1});

Medium – 120–280 kg K ha^{-1} (144–336 kg K_2O ha^{-1});

High – above 280 kg K ha^{-1} (336 kg K_2O ha^{-1}).

Based on exchangeable K about 21% soils are low, 51% medium and 28% high in available K in India (Hasan, 2002).

However, use of both EK and NEK has been suggested for deciding about K fertilization in India and GIS-based maps have been prepared to demarcate low- medium- and high-K districts (Srinivasarao *et al.*, 2010). Numbers of districts under different categories are given in Table 9.3. According to this categorization, 6% soils are in low-K category, 65% in medium-K category and 29% in high-K category.

Table 9.3: Characterization of Indian soils based on exchangeable and non-exchangeable potassium

Category	Exch. K	Non-exch K	No. of districts	Recommendation
I	Low	Low	15	K fertilization is a must
II	Low	Medium	18	K fertilization is essential
III	Low	High	2	K application at critical stages of crops improves yield
IV	Medium	Low	58	K application at critical stages
V	Medium	Medium	115	Maintenance dose in intensive cropping
VI	Medium	High	172	Crops may need immediate K application
VII	High	Low	1	Long-term cropping would need K after few years
VIII	High	High	24	K application is not required immediately
IX	High	High	129	K application not required

Limits for above categorization—Exchangeable K: Low < 50 mg K kg^{-1}, Medium 50–120 mg K kg^{-1}, High >120 mg K kg^{-1}; Non-exchangeable K: Low < 300 mg K kg^{-1}, Medium 300–600 mg K kg^{-1}, High > 600 mg K kg^{-1}.

Source: Srinivasarao *et al.* (2010)

POTASSIUM IN IRRIGATION WATER

Most irrigation water contains some K and this factor separates K management from N and P management. For example, an application of 1,000 mm water to rice in the Upper Gangetic Plains in India, containing about 5.9 mg K L^{-1}, would add 59 kg K ha^{-1}. However, in the Lower Gangetic Plains in Bengal, where the K content in irrigation water is low (about 1.02 mg K L^{-1}), the same amount of irrigation water would add only 10.2 kg K ha^{-1}. Similarly in the Cauvery Delta in Tamil Nadu, where irrigation water may contain 1.0 – 9.5 mg K L^{-1} will add 10–95 kg K ha^{-1} (Tiwari, 2014). Thus, irrigation water can meet 10 to 30% K needs of the crops and would tend to enrich the soil with K.

POTASSIUM FERTILIZERS

The most popular K fertilizer is potassium chloride or muriate of potash (MoP) containing 60% K_2O, making up about 90% or more of K consumed in India as a straight fertilizer or as a component of NPK fertilizers. Other K fertilizers are sulphate of potash (SoP) containing 48% K_2O, potassium magnesium sulphate and potassium nitrate. All K is imported in India. Potassium fertilizers are a non-renewable commodity and it has to be efficiently used. However, as of now there is not much concern about its getting in short supply as is the case with phosphate rock. The world deposits of K-minerals are estimated at 3,900 million metric tonnes (MMT). Leading K fertilizer-producing countries are Canada, Russia, Belarus, China and Germany (Table 9.4).

Table 9.4: Seven major K-fertilizer producing countries

Country	Production (MMT K_2O)	Reserves (MMT K_2O)
Canada	12.0	1,000
Russia	7.2	500
Belarus	6.4	750
China	6.2	360
Germany	2.9	150
Israel	2.2	270
Jordan	1.3	270
Total (World)	42.0	3,900

Source : US Geological Survey (2018)

POTASSIUM MANAGEMENT

Source

There is not much choice, since muriate of potash (MoP or KCl) is the main K fertilizer in India. However, potassium sulphate is the preferred K fertilizer for tobacco (*Nicotiana* sp.), since chlorine in KCl impairs the burning quality of flue-cured leaves used for cigarettes.

Rate of application

Rate of K application is to be guided by soil test for K and the likely response to K on a soil.

Time of application

In most crops, K is applied at planting. However, in potato (*Solanum tuberosum* L.), which is highly K-responsive crop, split application of K, 50 kg K_2O ha^{-1} at planting and 50 kg K_2O ha^{-1} at 45 days after planting is recommended from ICAR–Central Potato Research Institute, Shimla (Sud and Grewal, 1991). Adhikary and Karki (2006) from Nepal also made the same recommendation for potato.

Method of application

In most crops K is deep placed or broadcast and incorporated in soil at planting. In groundnut, Umar *et al.* (1999) recommended a foliar application of 1% KCl solution in addition to basal application of 50 kg K_2O ha^{-1} both for rainfed rainy season (*kharif*) and irrigated winter season (*rabi*) crops in Gujarat. Response to foliar K application was better in rainfed *kharif.* No injury to foliage was observed by foliar application of K. Sandhya Rani *et al.* (2014) from Nandyal, Andhra Pradesh recommended two sprays of 19-19-19 @ 2.5 kg ha^{-1} and top-dressing with MoP @ 80 kg K_2O ha^{-1} in addition to a basal application of 160 kg N + 80 kg P_2O_5 + 80 kg K_2O ha^{-1} in rice. Foliar application of K is also recommended for potato (Trehan *et al.*, 2008).

INTERACTIONS WITH OTHER NUTRIENTS

In normal rates of application, K interacts positively with most plant nutrients. However, since Ca, Mg and K occupy and compete with each other in the exchange complex of clay minerals and on absorption sites on plant roots, excessive amounts of any one of these is likely to interfere with the absorption of the other two. For example in barley, high K application rates reduced Ca absorption (Johanson *et al.*, 1968). Similarly, Ding *et al.* (2006) reported that Mg deficiency in rice was aggravated by heavy K fertilization. As regards, S and micronutrients, K is reported to have synergistic (positive) interaction with S, Zn, Mn and Cu and antagonistic (negative) interaction with B, Fe and Mo (Anonymous, 1998).

SUMMARY

1. Grain crops remove 30–50 kg K_2O t^{-1} grain.
2. Although removed in amounts larger than N and P, K concentration is the least in grains.
3. Most potassium is present in soil in inorganic form.
4. There are four forms of K in soil, namely, mineral–K (80–90%), Non-exchangeable–K (NEK) (3–9%), exchangeable–K or EK (0.5–3%) and solution K (< 0.5%).

5. NEK, EK and solution–K tend to maintain an equilibrium.
6. Available K in soil is estimated by extracting the soil with 1 N ammonium acetate.
7. Some researchers opine that fertilizer recommendations for K should take into account, both exchangeable–K and non-exchangeable–K.
8. Irrigation water can make a significant contribution towards meeting K needs of a crop; the contribution could range from 10 to 90 kg K_2O ha^{-1}, depending on the amount of irrigation water applied and its K concentration.
9. Split application of K is recommended in potato.
10. Foliar application of K has been recommended in potato and rice.
11. All K fertilizer is imported in India.
12. Canada is the largest producer of K fertilizers in the world.

REFERENCES

Adhikary, B.H. and Karki, K.B. 2006. Effect of potassium on potato tuber production in acid soils of Malepatan, Pokhra. *Nepal Journal of Agricultural Research* **7**: 42–48.

Annapurna, M.V.V.I., Kadam, B.S. and Surabhi, H.K. 2017. Distribution of different forms of potassium of representative soil series of sub-montane zone of Maharashtra, India. *International Journal of Current Microbiology and Applied Sciences* **6**(7): 2,648–2,653.

Anonymous. 1998. Potassium interactions with other nutrients. *Better Crops* **82**(3): 12–13.

Ding, Y., Luo, W. and Xu, G. 2006. Characterization of magnesium nutrition and interaction of magnesium and potassium in rice. *Annals of Applied Botany* **149**(2): 111–123.

Hasan, R. 2002. Potassium status of soils in India. *Better Crops International* **16**(2): 3–5.

Johanson, C., Edwards, D.G. and Loneragan, J.F. 1968. Interactions between potassium and calcium in their absorption by intact barley plants. *Plant Physiology* **43**: 1,717–1,721.

Mitra, G.N., Sarma, V.A.K. and Ramamoorthy, B. 1958. Comparative studies on the potassium fixation capacities of Indian soils. *Journal of the Indian Society of Soil Science* **6** : 1–6.

Muhr, G.R., Datta, N.P., Sankarasubramany, H., Dever, R.P., Leley, V.K. and Donahue, R.L. 1965. *Soil Testing in India*. United States Agency for International Development Mission in India. New Delhi, 120 pp.

Sandhya Rani, B., Krishna, G. and Munirathnam, P. 2014. Studies on the effect of foliar fertilization in combination with conventional fertilizers on yield, economics and nutrient uptake of rice (*Oryza sativa* L.) under KC Canal Ayacut area of Andhra Pradesh. *Agriculture Science Digest* **34**(1): 15–20.

Sekhon, G.S., Brar, M.S. and Subba Rao, A. 1992. *Potassium in some Benchmark Soils of India*. Potash Research Institute, Gurgaon, Haryana, India, 82 pp.

Shivay, Y.S, Prasad, R., Singh, R.K. and Pal, M. 2015. Relative efficiency of zinc-coated urea and soil and foliar application of zinc sulphate on yield, nitrogen, phosphorus, potassium, zinc and iron biofortification in grains and uptake by *Basmati* rice (*Oryza sativa* L). *Journal of Agricultural Science,* Canada **7**(2): 161–173.

Srinivasarao, Ch., Subba Rao, A., Rao, K.V., Venkateswarlu, B. and Singh, A.K. 2010. Categorization of districts based on non-exchangeable potassium: implication in efficient K fertility management in Indian agriculture. *Indian Journal of Fertilisers* **6**(7): 40–56.

Sud, K.C. and Grewal, J.S. 1991. Effect of time of potassium application on potato nutrition in Shivalik hills. *Journal of Potassium Research* **7**: 266–276.

Tiwari, K.N. 2014. Potassium management. (In) *Textbook of Plant Nutrient Management*, pp: 122–143. Prasad, R., Kumar, D., Rana, D.S., Shivay, Y.S. and Tewatia, R.K. (Eds.), Indian Society of Agronomy, New Delhi.

Trehan, S.P., Upadhyaya, N.C., Sud, K.C., Kumar, M., Jatav, M.K. and Lal, S.S. 2008. *Nutrient Management in Potato*. 72 pp. ICAR–Central Potato Research Institute, Shimla, Technical Bulletin 90.

Umar, S., Bansal, S.K., Imas, P. and Magen, H. 1999. Effect of foliar fertilization of potassium on yield, quality and nutrient uptake of groundnut. *Journal of Plant Nutrition* **22**(11): 1,785–1,795.

USGS. 2018. Potash Mineral Commodity Summary, US Geological Survey. January 2018 (via internet).

QUESTIONS

1. Briefly discuss the dynamics of K in soil.

2. Answer True (T) or False (F)

(a) K concentration in wheat grains is higher than P concentration.
(b) Germany is the largest producer of K fertilizers in the world.
(c) K by plant roots is mostly absorbed from solution and exchangeable K.
(d) Irrigation water contains no K.
(e) Split application of K is recommended for potato.

3. Match the following items

(a)	Non-exchangeable-K	(i)	High P-fixing soils
(b)	Mineral K	(ii)	Soils poor in K
(c)	Biotite	(iii)	8% of total K in soil
(d)	Vertisols	(iv)	80% of total K in soil
(e)	Red soils	(v)	Iron mica

Max Marks: 100 Sample Test Paper II (Major nutrients) Time: 2 Hrs

1. Briefly discuss (40 Marks)

(a) Nitrogen dynamics in soil
(b) Phosphate fixation in soil
(c) Environmental implications of fertilizer nitrogen.
(d) Potassium dynamics in soil.

2. Answer "True" or "False" (20 Marks)

(a) Organic N makes up about 30–40% of total N in soil.
(b) Bacteria *Nitrosomonas* sp. convert ammonium to nitrite.
(c) Exchangeable K constitutes less than 10% of available K in soils.
(d) DAP contributes about two-thirds of the total P consumption in India.

(e) Calcium ammonium nitrate contributes about 50% of total N consumption in India.
(f) Potassium sulphate is considered as a better K fertilizer than MoP for tobacco.
(g) For producing one tonne of wheat, the crop removes about 35 kg N.
(h) Fungi do not play a role in denitrification in soil.
(i) Non-exchangeable content in soil should also be considered along with exchangeable K in determining fertilizer rate on a soil.
(j) Ground phosphate rock can be used as a source of P for crops on acid soils.

3. Match the following (20 Marks)

(a) Variscite	(i) Iron mica
(b) Denitrifiaction	(ii) Sand particles
(c) Surface application of urea	(iii) $Ca(H_2PO_4)_2$
(d) Monocalcium phosphate	(iv) Reaction product of fertilizer P application
(e) Biotite	(v) Nitrous oxide
(f) Orthoclase	(vi) *Thiobacillus* sp
(g) Illite	(vii) $CaHPO_4$
(h) Nitrate pollution of drinking water	(viii) Ammonia volatilization
(i) Ozone depletion	(ix) Blue baby syndrome
(j) Dicalcium phosphate	(x) High K fixation

4. Fill up the blanks (20 Marks)

(a) For producing 1 tonne of rice, the crop removes .... kg N + ... kg P_2O_5 + ... K_2O.
(b) Ammonia volatilization losses are more on .... soils.
(c) Alternate wetting and drying leads to high .... losses.
(d) In addition to P, SSP also contributes .... to soil.
(e) Growing of a grain legume can contribute about .... kg N ha^{-1} to the succeeding crop in a cropping system.
(f) *Azotobacter* sp can fix about .... kg N ha^{-1}.
(g) DAP contains .... % N, ... % P_2O_5 and .... % K_2O.
(h) Acid soils damage the crops due to .... toxicity.
(i) Nitrogen from soils is leached as ....
(j) Neem-coated urea .... the efficiency of applied N.

Chapter 10

Secondary Nutrients

Secondary nutrients is a triad of three essential plant nutrients namely, Calcium (Ca), Magnesium (Mg) and Sulphur (S). Of these three, Ca and Mg are taken up by plants as cations, while S is taken up as anion SO_4^{2-}. Among the secondary plant nutrients, sulphur is emerging as a major plant-nutrient deficiency in crops grown in the Indo-Gangetic Plains, spread over 13 million ha in Pakistan, India, Nepal, and Bangladesh (Khurana *et al.*, 2008). From the Indian point of view, S deficiency has emerged in a large part of the country and needs to be attended. Uptake of Mg and S by some crops is given in Table 10.1.

Table 10.1: Average uptake of sulphur and magnesium by some crops in India

Crop	kg S t^{-1} grain/seed[1]	kg Mg t^{-1} grain/seed[2]
Cereals		
Rice	2.7	3.0
Wheat	3.9	4.7
Maize	2.4	7.8
Sorghum	5.4	4.8
Pearl millet	5.8	–
Pulses		
Pigeonpea	6.8	4.0
Chickpea	6.5	7.3
Blackgram (Urdbean)	9.2	–
Greengram (mungbean)	7.7	–
Oilseeds		
Rapeseed-mustard	15.3	8.6
Groundnuts (pods)	9.0	10.8
Sunflower	8.2	–
Safflower	10.8	–
Sesame	8.3	–
Soybean	6.3	–

Source: [1]Tandon (2011); [2]Shrotriya (2007)

CALCIUM AND MAGNESIUM IN SOIL

The Earth's crust contains 3.64% Ca and thus, soil as a medium of plant growth is more controlled by Ca than any other element. The Ca content in soils can vary from 0.1 to 0.3% in humid tropics to 0.7 to 1.5% in calcium carbonate-free soils of the humid temperate regions and < 1 to > 25% in calcareous soils of sub-humid sub-temperate regions (Jena and Rout, 2014), where it precipitates as $CaCO_3$ (Calcite) and is the main liming material. In several parts of the sub-humid sub-temperate regions of India, there are deposits of hard nodules of calcium carbonate with impurities of Mg, Fe and Al, known as kankar. The calcium content in kankar varies from 25.6 to 38.64% as CaO (Singh and Singh, 1972). Kankars are used for making kaccha (non-metalled) roads in the countryside. In Rajasthan desert areas, there are also large deposits of calcium sulphate ($CaSO_4$) or gypsum.

Magnesium content in the Earth's crust is 1.93%, nearly half that of Ca. Its concentration is also low in the acid soils of the humid tropic (~ 0.1%) and could be about 4% in the soils of the arid and semi-arid regions, where it is derived from Mg rich parent material (Jena and Rout, 2014). In nature it co-precipitates with Ca as dolomite ($CaCO_3MgCO_3$).

Being so abundant in Earth's crust and in surface soil, Ca and Mg dominate the exchange complex of soil clay minerals in neutral and alkaline soils and when they leach down, the exchange complex gets richer in H^+ ions and the soils become acidic in nature (Table 10.2). For good growth of crops, acid soils have to be limed. However, the major crop grown in acid soil regions in India is lowland rice and submergence raises soil solution *p*H (Ponnamperuma, 1972).

Nevertheless, for the second crop after rice, which in many areas is a pulse, lime ($CaCO_3$) application is likely to be beneficial. In regions, where parent material was rich in Na, leaching of Ca leads to the development of sodic soils, and gypsum ($CaSO_4.2H_2O$) has to be applied to replace Na in the exchange complex. Again, rice has been found to be the best crop on sodic soil, because submergpence lowers soil solution H^+. So, both in acid and alkaline (sodic) soils, a Ca compound ($CaCO_3$ or $CaSO_4$) is used as an amendment and this provides enough Ca to growing crop plants. For this reason, Ca as a plant nutrient has not received much importance, except in groundnut (*Arachis hypogaea* L.).

Calcium nutrition of groundnut

Flowers in the groundnut are formed in the lower part of the plant and after fertilization the gynophore grows horizontally with the small ovary at its tip, which penetrates the soil for its development. This stage in groundnut plant is known as 'pegging'. The developing pod cannot get enough Ca for development from the main plant and has to directly absorb it from the soil. Chahal and Virmani (1973) reported that when the labeled Ca was fed to the rooting zone, 66% of it was retained by the leaves and only 13.8% moved to the pods. However, when the labeled Ca was fed to the fruiting zone, 88.3% was retained by the pods and 11.7% moved to other parts of the plant. Therefore, application of gypsum to groundnut is recommended at pegging (Giri, 2002). In acid soils of *p*H 5.1, application of 1 t ha^{-1} of lime or 0.3 t ha^{-1} of gypsum increased the pod yield by 660 and 540 kg ha^{-1} respectively (Satyanarayan *et al.*, 1975). As regards soil fertility status, Ramachandrappa and Kulkarni (1992) reported that, 25% Ca saturation of the exchange complex of soil was enough for good pod development on a sandy-loam soil, while on a red sandy-soil, 75% Ca saturation of the soil-exchange complex was required. Some of the increase in yield in studies involving gypsum could be owing to response to S and this need to be sorted out by planning the treatments carefully.

Table 10.2: Ions present in the exchange complex of an acid, a calcareous and a sodic soil

Ion	Acid soil[1] (*p*H 4.8)	Calcareous soil[2] (*p*H > 8.0)	Sodic soil[2] (*p*H 8.5 – >9.0)
Ca (% of CEC)	9.3	62	33
Mg (% of CEC)	5.5	26	36
Na (% of CEC)	2.2	8	20
K (% of CEC)	1.1	4	2
H (% of CEC)	0.8	–	–
Al (% of CEC)	0.8	–	–

Source: [1]Dhanajaya (2003); [2]Jena and Rout (2014)

Calcium : Magnesium ratio

A Ca : Mg ratio in soil solution greater than 7 : 1 is considered undesirable for crop production and continuous liming of acid soils can create Mg deficiency. On the other hand, a Ca : Mg ratio less than 2 : 1 can cause Ca deficiency (Tisdale *et al.*, 1985).

Crop response to magnesium

Not many soils in India are deficient in Mg, and therefore this nutrient has not received much attention in India. This is because, a large part of cultivated soils have *p*H above 7 and are not acid. Similarly, in New Zealand also Mg deficiency is not a common occurrence; it is transient in nature and responses of cereals would not be expected unless soil levels are very low (Craighead, 2001; Gerendas and Fuhrs, 2013) although pastures respond good to Mg fertilization (Edmeades, 2004). However, response of crops to Mg fertilization is reported from several other countries. A meta-analysis of crop responses in China showed that yield increment under Mg fertilization was more remarkable and reliable in southern China than in northern China; and yield increment of fruit and vegetable crops with Mg fertilization was more substantial than cereal crops (Wang *et al.*, 2016). Similarly, good response of sugarbeet (*Beta vulgaris* L.) and potato to Mg fertilization is reported from Europe and the USA (Hossner and Doll, 1970; Draycott and Durrant, 1972).

SULPHUR

Sulphur being a component of amino acids methionine, cystine and cysteine and of vitamins thiamine (B1) and biotin (B7) is of great importance in plant, human, and animal nutrition (Prasad and Shivay, 2018). In plants, there are a number of other S-containing organic compound of importance including alliin (a derivative of cysteine and precursor of allicin, the compound responsible for pungency in garlic and onion), glucosinolates (pungent compound is members of *Brassicaceae* family, such as rapeseed-mustard, cabbage (*Brassica qleracea* var. *capitata* L.), broccoli (*Brassica qleracea* var. *italica* Plenck), and glutathione (an important antioxidant in plants and animals) (Prasad, 2014). It is interesting to note that like N, S expresses several oxidation states (valence), which are: –4 in pyrites, –2 in H_2S and FeS, +4 in SO_2 and +6 in H_2SO_4 or SO_4^{2-}.

Sulphur content in soil

Total S content in most Indian soils varies from 50 to 300 mg kg^{-1}, but may be as high as 1,000 mg kg^{-1} in saline sodic soils (Tewatia *et al.*, 2014).

Forms of Sulphur in soil

Since it is a part of proteins, S is closely associated with N and therefore organic-S makes about 90% of total S in soil. The global average N: S ratio in soil organic matter (SOM) is 7.22 : 0.75 (Tipping *et al.*, 2016). Thus, if SOM in a field releases 40 kg N ha^{-1}, it will release about 4.2 kg S ha^{-1}. Inorganic S is only about 10% of total S and is mostly present as sulphate (SO_4^{2-}), the form in which is taken up by the crop plants. Sulphate may be present in soil solution or it could be retained in the exchange complex of silicate minerals, especially kaolinite that have OH^- radicals on their edges or hydrous oxides of Al and Fe, which have a positive charge under acid conditions. Under anaerobic conditions, such as those obtained in lowland rice fields and in marshy lands it may occur as hydrogen sulfide (H_2S), which is a gas, that escapes to air and is responsible for the foul air in marshes. In lowland rice fields in acid regions, where ferrous (Fe^{2+}) may be present, H_2S can form FeS, responsible for the *Akiochi* disease of rice (Tanaka and Yoshida, 1970). In marshes, FeS can lead to the formation of pyrites (FeS_2).

Soil test for available sulphur

Available S in India is mostly determined by extracting soil S with 0.15% $CaCl_2$ solution (Williams and Steinberg, 1959), which would include solution-S, some exchangeable-S and some S extracted from organic matter. Critical level of available S is taken as 10 mg kg^{-1} soil. Based on this critical level, 42.3% of 135,000 soil samples from all over the country were found to be deficient in S. Sulphur deficiency is most in the western states, followed by the southern states and is the least in the north-eastern states of the country (Tewatia *et al.*, 2006). The matter of concern is the fact that the number of districts deficient in S increased from 70 in 1991 to 300 in 2011 (Tandon, 2011). The extent of S deficiency in different agro-ecological zones of India is shown in Fig. 10.1.

The main reasons for the increase in S deficiency over time in Indian soils are as under.

1. Growing of high-yielding varieties of crops, resulting in increased removal of S from soil.
2. Increase in cropping intensity, causing in increased removal of S from soil.
3. Shift from S-containing fertilizers, such as, SSP and ammonium sulphate to non-S-containing fertilizers, such as urea, DAP (diammonium phosphate), UAP (urea ammonium phosphate) etc.
4. Decreased use of organic manures, compost and crop residues.
5. Loss of S by erosion and leaching.

Sulphur management in crop production

Sources of sulphur: Source of S could be organic, straight S materials, S-containing conventional fertilizers, S-containing customized fertilizers and S containing micronutrient fertilizers. Common S sources are listed in Table 10.3.

Sulphur-coated urea (SCU) has been tested mostly as a slow-release nitrogen fertilizer. However, Shivay *et al.* (2016b) have reported that 5% S-coated urea increased N uptake by 24% and S uptake by 29% over prilled urea; the increase in N-recovery efficiency with SCU was about 60% over prilled urea. Hence, SCU holds promise as a source of S.

There are two important points that need to be discussed in connection with the sources of S.

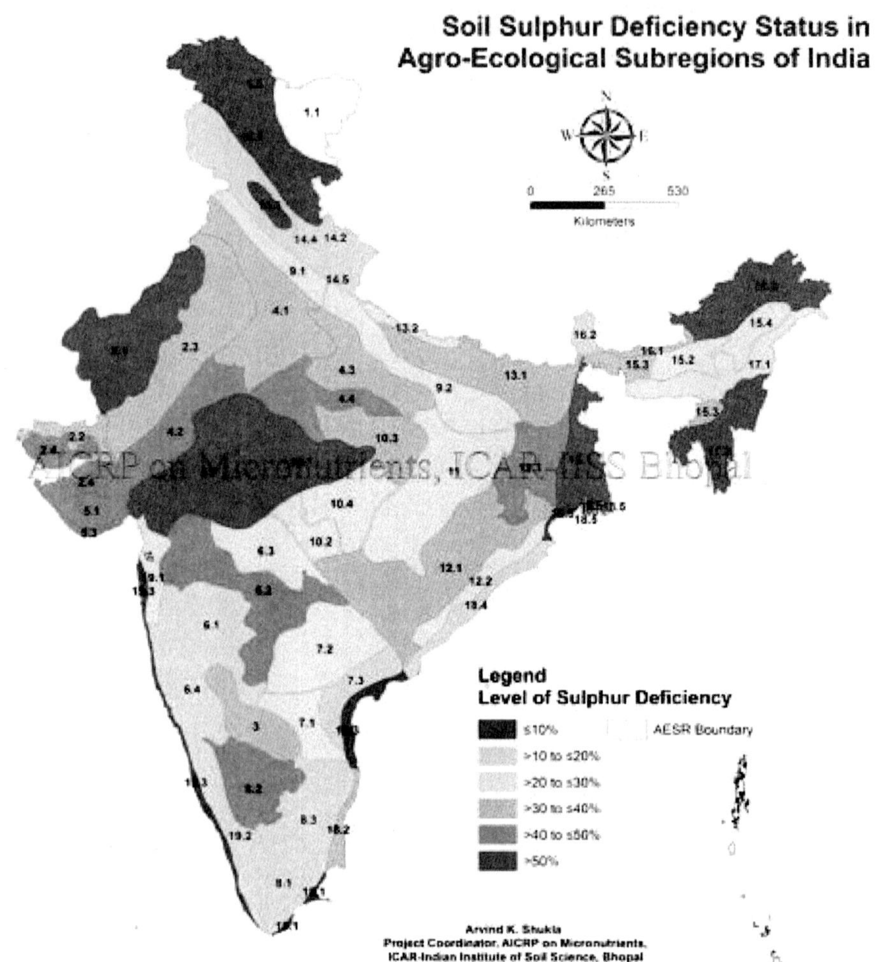

Fig. 10.1. Sulphur deficiency in different agro-ecological zones of India (with permission from the Project Coordinator, AICRP on Micronutrients, Indian Institute of Soil Science, Bhopal)

1. When elemental-S or bentonite-S or pyrite is used as a source of S, it has to be applied 3 to 4 weeks before sowing of a crop, because it has to be oxidized to SO_4^{2-} form, in which it is taken up by a plant. This oxidation is carried out by bacteria *Acidithiobacillus thiooxidans* (Waksman and Joffe) Kelly and Wood, *Thiobacillus thioparus Beijarinck, Acidithiobacillus ferrooxidans* (Temple and Colmer, Kelly and Wood etc.) (Keunen, 1975). These bacteria derive their energy from the oxidation of S to SO_4^{2-}. The general equation is as under:

Table 10.3: Sources of sulphur for crop production

Source	% S	Other nutrient(s)	Remarks
Organic			
Pressmud	2.3		Obtained from sugar factories
Straight S materials			
Elemental S	99	–	To be applied 3–4 weeks before sowing
Bentonite S	90	–	To be applied 3–4 weeks before sowing
Pyrites	22–24	20–22% Fe	To be applied 3–4 weeks before sowing
Gypsum	13–14	16–19% Ca	Good for groundnut
Phosphogypsum	16–20	23% Ca	From phosphoric acid factories
Conventional fertilizers			
Single superphosphate	11	16% P_2O_5	Good basal fertilizer for groundnut
Ammonium sulphate	23	20.6% N	Suitable for top dressing for S
Ammonium phosphate sulphate	13	16% N + 20% P_2O_5 20% N + 20% P_2O_5	Combined P and S application
Potassium sulphate	17.5	50% K_2O	K + S application especially in tobacco
NPKS 15–15–15–09	9	15% N + 15% P_2O_5 + 15% K_2O	Balanced NPK + S fertilization
NPKS 13–33–0–06	6	13% N + 33% P_2O_5	Combined P and S application
Micronutrient fertilizers			
Zn sulphate monohydrate	15	33% Zn	
Zn sulphate heptahydrate	10	21% Zn	
Ferrous sulphate	10.5	19% Fe	
Copper sulphate	12	24% Cu	
Manganese sulphate	17	30.5% Mn	
Others			
Magnesium sulphate	12	9.6% Mg	
Customized fertilizers	3–6	NPKZnB varying	Site-specific nutrient management

Source: Tewatia *et al.* (2014)

$$CO_2 + S + \frac{1}{2} O_2 + 2H_2O \longrightarrow [CH_2O] + SO_4^{2-} + 2H^+$$

The soil conditions that control the above reaction are *p*H, soil aeration, soil moisture and temperature. The above reaction is best under acid conditions with *p*H <5, and is contradictory to nitrifying bacteria, which require a *p*H value of 7 or above. The reaction is rapid in the temperature range of 20–40°C. Aerobic conditions are necessary for *Thiobacilli*. However, some oxidation can take place under anaerobic conditions by photosynthetic green S bacteria *Chlorobium* and purple S bacteria *Chromatium*, *Thiocystis* and *Thicapsa* (Fenchel and Blackburn, 1979). For more information; reference may be made to Prasad and Power (1997). *Acidithiobacillus ferrooxidans* is also responsible for oxidizing pyrite (Fenchel and Blackburn, 1979) and the general equation is as under:

$$2FeS_2 + 2H_2O + 7O_2 \longrightarrow 2Fe^{2+} + 4SO_4^{2-} + 4H^+$$

2. There have been a large number of experiments in India on response of crops to an application of zinc sulphate, but no efforts have been made in these studies to sort out Zn and S effects.

Table 10.4: Direct and residual response to sulphur in a cropping system

Crop receiving S	Succeeding crop	Response (kg ha^{-1} year^{-1})	Direct (%)	Residual (%)
Rice	Mustard	1,066	82	18
Groundnut	Wheat	1,186	56	44
Groundnut	Groundnut	299	47	53
Wheat	Groundnut	933	80	20

Source: Biswas and Tewatia (1991)

Rates of application: Recommendations for S fertilization are made on the basis of available S in soil and vary from 20 to 40 kg S ha^{-1}; higher amounts are required in oilseed crops, especially, rapeseed–mustard. The agronomic efficiency of S (AEs) or response to S application (kg grain or seed increased kg^{-1} S) is 22 for cereals, 12 for oilseeds and 7 for pulses (Tandon, 2011). However, one should not go by these numbers only, because the increase in seed yields in oilseeds and pulses though much lower than in cereals, fetch much higher prices than cereals. Sulphur fertilization to one crop in a cropping system leaves a significant residual effect on the succeeding crop. Some data are in Table 10.4.

Time and method of application: Sulphur deficiency is not easily visible in crops and the crops may suffer due to hidden hunger. It is therefore best applied at sowing as gypsum, phosphogypsum, SSP, zinc sulphate (if Zn is also to be applied), APS or any complex fertilizer-containing S. For top-dressing ammonium sulphate is the best choice, because N is generally top-dressed.

Sulphur and crop quality

1. Sulphur fertilization not only increases seed yield but also oil content in oilseeds (Aulakh and Pasricha, 1997; Giri, 2002).
2. Sulphur fertilization increased S-containing amino acids (SAAs) methionine content by 35.3% and cysteine content by 24.5% in winter wheat (Jarvan *et al.*, 2008). This is important, because SAAs are

important for human nutrition and the recommended dietary allowance for SAAs for adults is 13 mg kg^{-1} body weight (780 g day^{-1} for a person weighing 60 kg) (Parcell, 2002).

3. Sulphur fertilization increases S content in wheat grain and lowers N : S ratio, which improves baking quality of wheat flour (Prasad and Shivay, 2017).
4. Sulphur fertilization increases alliin content in garlic (*Allium sativum* L.) and onion (*Allium cepa* L.) (Bloem *et al.*, 2005). On cutting onion or garlic alliin produces allicin, which is responsible for their pungency (Prasad, 2014).
5. Sulphur fertilization enhances glucosinolate (GSL) content in cabbage (Beecher, 1994). GSLs can help in fighting against cancer (Das *et al.*, 2000).

Interactions with other nutrients

Sulphur reacts positively with N and B. Dubey and Khan (1993) showed that, response of mustard to S increased with an increase in the level of N and vice-versa (Table 10.5). When no N was applied, application of 30 kg S ha^{-1} resulted in 0.67 t ha^{-1} seed, while with 90 kg N ha^{-1} it resulted in 1.65 t ha^{-1} seed. Similarly, when no S was applied an application of 90 kg N ha^{-1} resulted in seed yield of 1.22 t ha^{-1}, while with 30 kg S ha^{-1}, it was 1.65 t ha^{-1}. The response to 30 kg S ha^{-1} was 0.22 t ha^{-1} in the absence of N and increased to 0.43 t ha^{-1} with 90 kg N ha^{-1}. Again, the response to 90 kg N ha^{-1} was 0.77 t ha^{-1} in the absence of S and 0.98 t ha^{-1} in the presence of 30 kg S ha^{-1}

Table 10.5: Seed yield of mustard as influenced by nitrogen and sulphur fertilization

Nitrogen (kg ha^{-1})	Sulphur (0 kg ha^{-1})	Sulphur (30 kg ha^{-1})	Response to N
0	0.45 t ha^{-1}	0.67 t ha^{-1}	0.22
90	1.22 t ha^{-1}	1.65 t ha^{-1}	0.43
Response to N	0.77 t ha^{-1}	0.98 t ha^{-1}	
CD (P=0.05)		0.11	

Source: Dubey and Khan (1993)

Similarly, Karle and Babula (1985) showed that, S fertilization increased the oil yield in groundnut by 73% and B fertilization increased it by 29%, while the combined application of S and B increased it by 181%.

Sulphur also interacts positively with Fe, Mn, Zn and Cu and sulphur fertilization as SCU increased the concentration of these micronutrients in wheat grain (Shivay *et al.*, 2016a) (Table 7.3). This could be owing to increased availability and absorption of these micronutrients, because of creation of increased acid conditions by SO_4^{2-} ions in the rhizosphere.

Table 10.6: Effect of nitrogen (N) and Sulphur (S) fertilization on N and S concentration and N : S ratio in wheat grain

kg N + kg S ha^{-1}	% N in grain	% S in grain	N : S ratio
0 + 0 (Control)	1.65	0.18	9.2
130 + 0	1.85	0.19	9.7
130 + 15.70	2.07	0.22	9.4
LSD (P=0.05)	0.006	0.02	0.6

Source: Shivay *et al.* (2016b)

However, since proteins contain S-amino acids methionine, cystine and cysteine, S fertilization increases their content and this may decrease N : S ratio in grain, which is considered desirable, since low N : S ratio in wheat flour is associated with high loaf height and volume in bread (Unger *et al.*, 2002). Some data on the effect of S fertilization on N : S ratios in wheat are given in Table 10.6.

SUMMARY

1. Magnesium (Mg) removal by rice is about 3 kg t ha^{-1}.
2. Calcium (Ca) saturation in soil-exchangeable complex is about twice that of Mg.
3. In sodic soils, Na saturation of soil-exchange complex is 15% or higher.
4. A Ca : Mg ratio larger than 7 : 1 is considered undesirable for plant growth.
5. Continuous liming at high rates can lead to Mg deficiency.
6. The Mg deficiency occurs in highly leached acid soils.
7. Sulphur (S) is a constituent of amino acids methionine, cystine and cysteine and a number of other compounds in plants such as, glutathione, glucosinolates and alliin.
8. Total S content in soils can vary from 50 to 300 mg kg^{-1}.
9. About 90% S is present as organic S and N : S ratio in SOM is about 7 : 1.
10. Inorganic S is present as SO_4^{2-} in soil or in soil exchange complex.
11. Available S in soil is determined by extraction with 0.15% $CaCl_2$ solution.
12. Straight S-supplying materials include elemental S, bentonite S, pyrites, gypsum and phosphogypsum.
13. When elemental S, bentonite S or pyrites are used a source of S, they should be supplied 3–4 weeks before sowing and soil should be kept moist, but not flooded.
14. Single superphosphate (SSP) is a good source of S for basal fertilization, while ammonium sulphate is ideal for top-dressing S.

15. It is always better to apply S as basal and not wait for S deficiency in plants to show up, because if S is not applied the crop may suffer due to hidden hunger.
16. Elemental S and bentonite are oxidized by *Acidithiobacillus thiooxidans*, *Thiobacillus thioparus* and *Acidithiobacillus ferrooxidans*. These bacteria function well under acid conditions.
17. Pyrite is oxidized by *Acidithiobacillus ferrooxidans* only.
18. Photosynthetic green S bacteria *Chlorobium* and purple S bacteria *Chromatium*, *Thiocystis* and *Thicapsa* can oxidize S under reduced conditions.
19. Sulphur (S) requirement of oilseeds, especially rapeseed–mustard is much higher than that of cereals.
20. Sulphur (S) fertilization increases oil content in oilseeds, SAAs in cereals and alliine in garlic and onion.
21. Sulphur (S) interacts positively with N and B.

REFERENCES

Aulakh, M.S. and Pasricha, N.S.1997. Role of balanced fertilization in oilseed-based cropping system. *Fertiliser News* **42**(40): 101–111.

Beecher, C. 1994. Cancer preventive properties of *Brassica oleracea*: A review. *American Journal of Clinical Nutrition* **59**: 1,166–1,170.

Biswas, B.C. and Tewatia, R.K. 1991. Results of FAO sulphur trials network in India. *Fertiliser News* **36**(4): 11–35.

Bloem, E., Haneklauss, S. and Schnug, E. 2005. Influence of nitrogen and sulphur fertilization on alliin contents of onions and garlics. *Journal of Plant Nutrition* **27**: 1,827–1,839.

Chahal, R.S. and Virmani, S.M. 1973. Uptake and translocation of nutrients in groundnut. I. *Oleagineux* **80**:181–184.

Craighead, M. 2001. Magnesium deficiency in crops and its relevance to arable farming in New Zealand – a review. *Agronomy New Zealand* **31**: 53–62.

Das, S., Tyagi, A.K. and Kaur, H. 2000. Cancer modulation by glucosinolates: a review. *Current Science* **79**:1,665–1,670.

Dhanajaya, B.C. 2003. Cation exchange equilibria in acid soils of Karnataka. Ph.D. Thesis, University of Agricultural Sciences, Bengaluru, Karnataka.

Draycott, A.P. and Durrant, M.J. 1972. Comparisons of kieserite and calcined magnesite for sugar beet grown on sandy soils. *Journal of Agricultural Science* **79**: 455–461.

Dubey, O.P. and Khan, R.A. 1993. Effect of nitrogen and sulphur on dry matter, grain yield and nutrient content at different growth stages of mustard (*Brassica juncea* L.) under irrigated Vertisol. *Indian Journal of Agronomy* **38**: 270–276.

Edmeades, C.D. 2004. The magnesium requirements of pastures in New Zealand: A review. *New Zealand Journal of Agricultural Research* **47**(3): 363–380.

Frenchel, T. and Blackburn, T.H. 1979. *Bacteria and Mineral Cycling*. 225 pp. Academic Press, New York.

Gerendas, J. and Fuhrs, H. 2013. The significance of magnesium for crop quality. *Plant and Soil* **368**: 101–128.

Giri, G. 2002. Groundnut, pp: 397–434. (In) *Textbook of Field Crops Production*. Prasad R. (Ed.). Indian Council of Agricultural Research, New Delhi.

Hossner, L.R. and Doll, E.C. 1970. Magnesium fertilisation of potatoes as related to liming and potassium. *Soil Science Society of America Proceedings* **34**: 772–774.

Jarvan, M., Eldesi, L., Adamson, A., Lukme, L. and Akk, A. 2008. The effect of sulphur fertilization on yield, quality of protein and baking qualities of wheat. *Agronomy Research* **6**: 459–469.

Jena, D. and Rout, K.K. 2014. Calcium and magnesium management. (In) *Textbook of Plant Nutrient Management*, pp: 156–181. Prasad, R., Kumar, D., Rana, D.S., Shivay, Y.S. and Tewatia, R.K. (Eds.), Indian Society of Agronomy, New Delhi.

Karle, B.G. and Babula, A.V. 1985. Effect of boron and sulphur on yield attributes and quality of groundnut. *Proceedings TNAU-FACT Seminar on Sulphur*, Coimbatore, pp. 158–168.

Keunen, J.G. 1975. Colourless sulphur bacteria and their role in sulphur cycle. *Plant and Soil* **43**: 49–76.

Khurana, M.P.S., Sadana, U.S. and Bijay-Singh. 2008. Sulfur nutrition of crops in the Indo-Gangetic plains of South Asia. (In) *Sulfur- A Missing Link between Soils, Crops, and Nutrition*. Agronomy Monograph 50. Jez, J. (Ed.), American Society of Agronomy, Crop Science Society of America, Soil Science Society of America, Madison, WI, USA, pp. 11–24.

Parcell, S. 2002. Sulphur in human nutrition and application in medicine. *Alternative Medicine Review* **7**: 22–44.

Ponnamperuma, F.N. 1972. The chemistry of acid soils. *Advances in Agronomy* **24**: 29–96.

Prasad, R. 2014. Major sulphur compounds in plants and their role in human nutrition and health –An overview. *Proceedings of Indian National Science Academy* **80**(5): 1,045–1,054.

Prasad, R. and Power, J.F. 1997. *Soil Fertility Management for Sustainable Agriculture*. 356 pp. CRC Press, Boca Raton, FL, the USA.

Prasad, R. and Shivay, Y.S. 2017. Nitrogen and sulphur concentrations and their ratio in plant tissues and foods and their relevance to plant and human nutrition: an overview. *Indian Journal of Fertilisers* **13**(2): 20–31.

Prasad, R. and Shivay, Y.S. 2018. Sulphur in soil, plant and human nutrition. *Proceedings of National Academy of Sciences, India, Section B: Biological Sciences* **88**: 429–433.

Ramchandrappa, B.K. and Kulkarni, K.R. 1992. Response of groundnut to Ca saturation in two soils. *Journal of Oilseed Research* **9**: 51–58.

Satyanarayan, T., Bahadur, U.P. and Havanagi, G.V. 1975. Studies on the influence of Ca on ragi and groundnut. *Journal of the Indian Society of Soil Science* **28**: 271–274.

Shivay, Y.S., Pooniya, V., Prasad, R., Pal, M. and Bansal, R. 2016b. Sulphur-coated urea as a source of sulphur and an enhanced efficiency of nitrogen for spring wheat. *Cereal Research Communications* **44**(3): 1–11.

Shivay, Y.S., Prasad, R., Pooniya, V., Pal, M. and Bansal, R. 2016a. Effect of sulphur fertilization on biofortification of wheat (*Triticum aestivum*) grains with Fe, Zn, Mn and Cu. *Indian Journal of Agricultural Sciences* **86**(6): 823–824.

Shrotriya, G.C. 2007. Magnesium in balanced fertilization. *Indian Journal of Fertilisers* **3**(1): 83–86.

Singh, L. and Singh S. 1972. Chemical and morphological composition of Kankar nodules in soils of the Vindahyan region of Mirzapur, India. *Geoderma* **7**(3–4): 269–276.

Tanaka, A. and Yoshida, S . 1970. *Nutritional Disorders of Rice Plant in Asia. Technical Bulletin*, No. 10. International Rice Research Institute, Los Banos, Philippines.

Tandon, H.L.S. 2011. *Sulphur in Soils, Crops and Fertilizers*. Fertilizer Development and Consultation Organization, New Delhi, 204 pp.

Tewatia, R.K., Choudhary, R.S. and Kalwe, S.P. 2006. TSI-FAI-IFA sulphur project-salient findings. *Proceedings of the TSI-FAI-IFA Symposium-cum Workshop on Sulphur in Balanced Fertilization*, New Delhi, pp. 15–25.

Tewatia, R.K., Yadav, D.S. and Shivay, Y.S. 2014. Sulphur management (In) *Textbook of Plant Nutrient Management*, pp: 144–156. Prasad, R., Kumar, D., Rana, D.S., Shivay, Y.S. and Tewatia R.K. (Eds), Indian Society of Agronomy, New Delhi.

Tipping, E., Luster, J. and Somerville, C. 2016. The C: N: P: S stoichiometry of soil organic matter. *Geophysical Research Abstracts* **18**, EUG 2016–1238, 20216.

Tisdale, S.L., Nelson, W.L. and Beaton, J.D. 1985. *Soil Fertility and Fertilizers*, edn 4, Macmillan, New York, the USA.

Unger, C., Flatyen, D.N., Grant, C. and Lukow, O.M. 2002. Impact of sulphur fertilization on spring wheat bread making quality. *Proceedings of the Great Plains Soil Fertility Conference*, Denver, CO, USA, 5–6 March, 2002, pp: 107–120. (www.ppi.org).

Wang, Y-H., Li., Y-J., He, X-H., Karim, M.R., Shi, X-J., Xie, D-T. and Zhang, Y-Q. 2016. Response of crop yields to magnesium fertilization: A meta-analysis in China. *International Journal of Agricultural Sciences* **6**(4): 1,011–1,015.

Williams, C.H. and Steinberg, A. 1959. Soil sulphur fractions as chemical indices of available sulphur in some Australian soils. *Australian Journal of Agricultural Research* **10**: 340–352.

QUESTIONS

1. Write brief notes on the followings

(a) Sulphur and food quality
(b) Lime plastering
(c) Sulphur-containing micronutrient fertilizers, giving their micronutrient and S composition.
(d) Role of gypsum in reclaiming sodic soils.
(e) Magnesium deficiency in soils

2. Answer True (T) or False (F)

(a) Sodic soils are richer in Ca than calcareous soil.
(b) Liming increases the activity of *Thiobacilli*-oxidizing S in soil.
(c) There is no oxidation in submerged soils.
(d) Calcium requirement of groundnut is high, because the developing pods also need Ca.
(e) Sulphur is a constituent of edible oils and this is why S fertilizer increases oil content in oilseeds.
(f) Gypsum supplies S and Ca to the soil.
(g) Sulphur fertilization of a crop leaves some residual effect on the succeeding crop.
(h) N : S ratio in SOM is 10 : 1.

3. Match the followings

(a) % S in SSP	(i) 10
(b) % S in ammonium sulphate	(ii) 11
(c) Sulphur uptake by rice (kg t^{-1})	(iii) 23

(d)	Sulphur uptake by pigeonpea (kg t^{-1})	(iv)	13–14
(e)	% Sulphur in zinc sulphate heptahydrate	(v)	7.8
(f)	Magnesium uptake by maize (kg t^{-1})	(vi)	3.64
(g)	Sulphur uptake by rapeseed–mustard (kg ha^{-1})	(vii)	6.8
(h)	% Ca in earth's crust	(viii)	2.7
(i)	% S in gypsum	(ix)	15.3
(j)	% Mg in earth's crust	(x)	1.93

Chapter 11

Micronutrients

There are nine essential micronutrients, namely, iron (Fe), manganese (Mn), zinc (Zn), copper (Cu), boron (B), molybdenum (Mo), chlorine (Cl), vanadium (V) and nickel (Ni). Out of these, Fe, Mn, Zn, Cu and Ni (Dominic *et al.*, 1978) are taken up by plants in their ionic form X^{2+}, Mo as MoO^{2-}, chlorine as Cl^- form and boron as boric acid (H_3BO_3) or as its ionic form B $(OH)_4^-$. The absorption of vanadium by plants has received only limited attention (Lepp *et al.*, 1983); it may be absorbed as cation V or vanadyl (VO^{2+}) and anion vanadate (VO_4^{3-}). Bulk of the information is available on Fe, Mn, Zn and Cu, because of the acceptability of the single DTPA (Diethylene triamine pentaacetic acid)–$CaCl_2$ extraction test of Lindsay and Norvell (1978) for availability of all these four nutrients followed by their estimation on Atomic Absorption Spectrometer (AAS). Available B is estimated by hot 0.15 M $CaCl_2$ extraction method (Saha and Singh, 1997). Molybdenum data base is poor, because its determination is difficult and Mo deficiency in crops has not been much noticed in India. Chloride is available in plenty in irrigation water and soil in India and has not received much attention in India.

There is an important variation in grading soils on the basis of soil tests for an available micronutrient from a primary or secondary nutrient. In the case of primary and secondary nutrient, the soils are defined as low, medium and high on the basis of a value for an available soil test. However, in the case of micronutrients, the soils are defined as deficient below a specified level, which is known as *Critical Deficiency Level* (*CDL*). So in the case of micronutrients, there are only two classes, namely deficient and sufficient.

The data on the extent of Zn, Cu, Fe, Mn and B deficiency in different states of India, based on the percentage of soil samples analyzing below critical level are given in Table 11.1

General critical deficiency values for different soils and crops in India for the DTPA-extractable micronutrients are: Zn 0.5–1.0 mg kg^{-1} soil, Cu 0.5–1.0 mg kg^{-1} soil, Fe 2.5–5.8 mg kg^{-1} soil, Mn 1.0–2.0 mg kg^{-1} soil, and B 0.5–1.0 mg kg^{-1} soil.

Of the five micronutrients for which data are provided in Table 11.1, Zn and B deficiency is wide-spread in the country and therefore separate chapters are provided for their agronomic management.

Table 11.1: Percent soil samples deficient in Zn, Cu, Fe, Mn and B in different states of India

Region–State	Zn[1]	Cu[1]	Fe[1]	Mn[1]	B[2]
Eastern					
Asom	34	<1	2	–	39
Bengal	36	<1	0	3	40
Bihar	54	3	6	2	52
Odisha	52	<1	0	0	52
Meghalaya	57	2	0	3	58
Central					
Uttar Pradesh	46	1	6	3	48
Madhya Pradesh	44	>1	7	1	47
North-western					
Punjab	46	<1	14	0	49
Haryana	60	2	20	4	60
Himachal Pradesh	42	0	27	5	NA
Western					
Gujarat	24	4	8	4	25
Maharashtra	83	1	24	1	NA
Rajasthan	21				
Southern					
Andhra Pradesh	47	<1	3	1	50
Karnataka	73	5	35	17	78
Tamil Nadu	59	6	17	6	59
Kerala	34	31	<1	0	35
All states	48	3	12	5	42

Source: [1]Singh (2009); [2]Singh (1995).

A brief discussion on other micronutrients follows:

IRON

Content and forms in soil

Iron is abundant in soils and occurs in concentrations of 7,000 to 500,000 mg kg^{-1} in soils (Fageria *et al.*, 2002). It is mostly present in insoluble Fe^{3+} (ferric) form, which hydrolyses readily to Fe $(OH)_2$, Fe $(OH)_3$, Fe $(OH)_4^-$ forms. It is the Fe^{2+} (ferrous) form, which is soluble in water and is taken up by plants. The solubility of Fe decreases 1,000-fold for each unit increase in soil *p*H in the range of 4–9 (Lindsay, 1979). Thus, iron deficiency occurs on calcareous soils having *p*H above 8.0.

Absorption of iron in plants

Organic matter plays an active role in increasing Fe availability by producing organic acids, such as, citric, malic, phenolic, which form complexes with Fe and make it possible for plant roots to absorb it. Plants have devised several mechanisms for absorbing the needed amounts of iron. This is achieved by different ways in non-graminaceous (beans, *Brassicas* etc.) and graminaceous (cereals, grasses etc.) plants. In non-graminacous plants, the Fe-acquisition mechanism involves acidification and reduction of Fe^{3+} to Fe^{2+} (Bauer and Hell, 2006), while in graminaceous plants it is achieved by the production of Fe-siderophores (FePSP) by plant roots (Ishimaru *et al.*, 2006).

Iron Fertilizers

Ferrous sulphate ($FeSO_4.7H_2O$) containing 18–20% Fe is the most popular Fe fertilizer and is to be generally applied on foliage. Chelated Fe fertilizer NaFeEDTA (5–14% Fe) is another fertilizer, preferred by some for foliar application; it is much costlier than ferrous sulphate. Organically complex Fe fertilizers, such as Fe-lignosulphate (5–8% Fe) are also available. Foliar application is a better method for Fe fertilization (Yadav *et al.,* 2013; Shivay and Mandi, 2020).

Iron deficiency in India

Iron deficiency has been reported in sorghum, sugarcane, groundnut and pulses in Tamil Nadu and Maharashtra on calcareous soils having *p*H above 8.0 (Kannan, 1984). In extreme Fe deficiency, leaves appear pale or may be completely bleached. The general method for correcting iron chlorosis has been the application of Fe salts to soil or as foliar sprays; the latter is a better method. In Tamil Nadu, an organically complexed Fe was found to increase crop yield in sorghum (Kannan, 1984). Singh and Dayal (1992) reported that, the maximum increase (16 to 24%) in pod yield of groundnut was recorded with a spray of iron sulphate + citric acid, while yield increases were 13 to 19, 12 to 18, and 11 to 18% over the control with iron sulphate, iron citrate and Fe–EDTA sprays, respectively. Splitting the application dose of 2 kg Fe ha^{-1} in two sprays @ 1 kg Fe ha^{-1} each at 50 and 70, days after emergence (DAE) or in four sprays at 0.5 kg each at 30, 45, 60, and 75 DAE, resulted in higher recovery from chlorosis, and in the highest pod and haulm yields. Manjappa and Rao (2001) from Raichur (Karnataka) reported more bolls/plant and heavier bolls and increased seed cotton yield with 0.5% spray of ferrous sulphate.

Considering wide-spread Fe deficiency anemia in humans in India, one or two foliar applications of Fe, irrespective of yield response, were recommended for biofortfication of cereal grains with Fe, since the cereals are staple food in India (Prasad and Shivay, 2018).

MANGANESE

Mn deficiency was reported in wheat on sandy soils in Punjab by Takkar *et al.* (1997), but it was sporadic. Mn deficiency is not a problem in India at this moment. On the other hand Mn-Fe toxicities are likely to occur in submerged lowland rice fields and the symptoms are interveinal brown spots, stunted plants and reduced tillering (Doberman and Fairhurst, 2000).

COPPER

From a limited number of on-farm trials in Bihar, Haryana, Punjab and Tamil Nadu, Shukla and Behra (2012) reported an increase of 0–1.78 t ha^{-1} in cereals, 0.2–0.3 t ha^{-1} in millets, 0–0.8 t ha^{-1} in oilseeds, 0.3–0.5 t ha^{-1} in sugarcane and 4.4–6.2 t ha^{-1} in onions. Kumar *et al.* (2010) also reported a significant response of barley to copper fertilization in Doon-Valley in Uttarakhand. Silviya and Stalin (2017) conducted field experiments on 12 locations, varying in available Cu from 0.33 to 2.12 mg Cu kg^{-1} soil in the irrigated rice tract of Tamil Nadu with graded levels of Cu (0, 0.5, 1.00, 1.50, 2.00, 2.50 kg ha^{-1}) and found that application of Cu @ 1.5 kg ha^{-1} recorded the highest plant height (95.0 cm), number of productive tillers m^{-2} (323.6), number of filled grains $panicle^{-1}$ (107.6), test grain weight (24.5 g), grain yield (6.50 t ha^{-1}) and straw yield (7.74 t ha^{-1}). Further, the highest values were recorded in the location No. 4 where the initial soil Cu content was minimum (0.84 mg Cu kg^{-1} soil), which is slightly above the general critical deficiency level of 0.2–0.5 mg Cu kg^{-1} soil. However, in general Cu deficiency is not a major issue in India.

MOLYBDENUM

Anderson *et al.* (1942) were the first to report the response of a subterranean clover (*Trifolium subterraneum* L.) – perennial ryegrass (*Lolium perenne* L.) – harding grass (*Phalaris aquatica* L.) pasture to an application of 1 kg ha^{-1} of ammonium molybdate from South Australia. Reports of response of wheat and oats were later reported from Western Australia (Gartell, 1966). Takkar *et al.* (1989) reported response of rice, wheat, soybean, greengram, sorghum, pearlmillet [*Cenchrus americanus* (L.) Morrone], peas (*Pisum satisvum* L.) and groundnut to Mo application in India. Velmurugan *et al.* (2013) carried out a soil survey in soils of Madurai and Sivagangai districts of Tamil Nadu. A total of 202 surface soil samples were collected and analyzed for available Mo, which was found to vary from 0.028 to 0.961 mg kg^{-1} soil. The critical limit of available Mo was determined to be 0.043 mg kg^{-1} soil for greengram [*Vigna radiata* (L.) R. Wilczek]. Based on this critical limit, the soils were classified into three categories: (*i*) low: $<$ 0.043 mg kg^{-1} soil; (*ii*) medium: 0.043—0.082 mg kg^{-1} soil; (*iii*) high: $>$ 0.082 mg kg^{-1} soil. Overall, 3–41%

area in Madurai and 7– 46% of area in Sivagangai districts was found to be in the low to medium category in respect of available Mo. Greengram responded highly to Mo application in soils below the critical limit, whereas in soils with Mo greater than 0.082 mg kg^{-1} soil there was no response. Application of 0.075 mg of Mo kg^{-1} soil or 0.4 kg ha^{-1} as sodium molybdate was sufficient to optimize greengram yield in the major soil series of the districts. Nevertheless, Mo deficiency is not a major problem in India at the moment.

CHLORINE

Chlorine is abundant in soils and its concentration in soil solution may vary from 0.5 to 6×10^3 mg kg^{-1} soil; Cl^{-1} is often the dominant anion in water in saline soils. Coconut (*Coccos nucifera* L.) and palm (*Areca* sp.) oil require high amounts of chloride and may show its deficiency, when grown 20–25 km away from the sea and may respond to chloride fertilization (von Uexkull and Sanders, 1986; Ollagnier and Olivin, 1984).

NICKEL AND VANADIUM

Need for Ni fertilization in soybean, chickpea and temperate cereals have been reported (Mishra and Kar, 1974; Brown *et al*., 1987). As regards, V, its essentiality has been reported only for certain microorganisms, such as, *Anabaena variabilis* (Yakumin *et al*., 1991). No reports on these nutrients are available from India.

SUMMARY

1. Out of the nine micronutrients, the deficiency of Zn and B is wide spread in India.
2. Iron deficiency is reported from Karnataka and Tamil Nadu on high *p*H calcareous soils.
3. Foliar application of Fe should be preferred to soil application.
4. Deficiencies of Mn, Cu and Mo are sporadic.
5. Nickel and vanadium have still not been tested in India.

REFERENCES

Anderson, A.J. 1942. Molybdenum deficiency on a south Australian ironstone soil. *Journal of Australian Institute of Agricultural Sciences* **8**: 73–75.

Bauer, P. and Hell, R. 2006. Translocation of Fe in plant tissue. (In) *Iron nutrition in Plants and Rhizosphere Microorganisms*, pp. 279–288. Barton L.L. and Abadia J. (Eds.) Springer, Dordrecht, Netherlands.

Brown, P.H., Welch, R.M. and Cary, E.E. 1987. Nickel: a micronutrient essential for higher plants. *Plant Physiology* **85**: 801–803.

Doberman, A. and Fairhurst, T. 2000. *Rice Nutrient Disorders and Nutrient Management*. PPIC of Canada and International Rice Research Institute, Los Banos, Philippines. 191 p.

Dominic, A.C., Garland, T.R. and Wildung, R.E. 1978. Nickel in plants. I. Uptake kinetics using intact soybean seedlings. *Plant Physiology* **62**: 563–565.

Fageria, N.K., Baligar, V.C. and Clark, R.B. 2002. Micronutrients in crop production. *Advances in Agronomy* **50**: 107–150.

Gartell, J.W. 1966. Field response of cereals to molybdenum. *Nature* (London) **209**: 1050.

Ishimaru, Y., Suzuki, M., Tsukamoto, T., Suzuki, K., Nakazono, M., Kobayashi, T., Wada, Y., Watanabe, S., Matsuhashi, S., Takahashi, M., Nakanishi, H., Mori, S. and Nishizawa, N.K. 2006. Rice plants take up iron as Fe($^{3+}$)-phytosiderophore and as Fe($^{2+}$). *Plant Journal* **45**: 335–346.

Kannan, S. 1984. Problems of iron deficiency in different crop plants in India: Causative factors and control measures. *Journal of Plant Nutrition* **7**(1–5): 187–200.

Kumar, P., Kumar, S., Singh, D.K. and Singh, A. 2010. Response of barley to copper fertilization in acid soils of Doon Valley. *Indian Journal of Fertilisers* **6**(7): 66–69.

Lepp, N.W., Morrell, B.G. and Phipps, D.A. 1983. Vanadium absorption by plants: the uptake of vanadium by excised barley roots (*Hordeum vulgare cv Maris Mink*). *Inorganica Chimica Acta* **79**: 228–229.

Lindsay, W.I. 1979. *Chemical Equilibria in Soils*. Wiley, New York, USA.

Lindsay, W.I. and Norvell, W.A. 1978. Development of DTPA soil test for zinc, iron, manganese and copper. *Soil Science Society of America Journal* **42**: 421–448.

Manjappa, K. and Rao, S. 2001. Response of cotton to soil and foliar application of secondary and micronutrients and irrigated conditions. *Journal of Cotton Research and Development* **15**(2): 233–234.

Mishra, D. and Kar, M. 1974. Nickel in plant growth and metabolism. *Botanical Review* **40**: 395–449.

Olangier, M. and Olivin, J. 1984. Effects of potassium nutrition on the productivity of the oil palm. *Oleagineux* **39**: 349–363.

Prasad, R. and Shivay, Y.S. 2018. Iron fertilization of cereal crops for human health – A new concept. *Indian Journal of Fertilisers* **14**(5): 68–77.

Saha, J.K. and Singh, M.V. 1997. Boron adsorption-desorption characteristics of some major soil groups of India. *Journal of the Indian Society of Soil Science* **45**: 271–274.

Shivay, Y.S. and Mandi, S. 2020. Advances in understanding iron cycling in soils, uptake/use by plants and ways of optimising iron-use efficiency in crop production. (In) *Achieving Sustainable Crop Nutrition*, Rengel Z. (Ed.), Burleigh Dodds Science Publishing, Cambridge, UK, 2020. (ISBN: 9781 78676 312 9).

Shukla, A.K. and Behera, S.K. 2012. Micronutrient fertilizers and higher productivity. *Indian Journal of Fertilisers* **8**(4): 100–117.

Silviya, R.A. and Stalin, P. 2017. Rice crop response to applied copper under varying soil available copper status at Tamil Nadu, India. *International Journal of Current Microbiology and Applied Sciences* **6**(8): 1,400–1,408.

Singh, A.L. and Dayal, D. 1992. Foliar application of iron for recovering groundnut plants from lime induced iron deficiency chlorosis and accompanying losses in yields. *Journal of Plant Nutrition* **15**(9): 1,421–1,433.

Singh, M.V. 1995. Boron in soils and crops. (In) *Progress Report AICARP on Micronutrients*, Indian Institute of Soil Science, Bhopal, Madhya Pradesh, pp. 1–86.

Singh, M.V. 2009. Micronutrient nutritional problems in India for human and animal health. *Indian Journal of Fettilisers* **5**(4): 11–56.

Takkar, P.N., Chhiba, I.M. and Mehta, S.K. 1989. Twenty years of coordinated research on micronutrients in soils and plants. Bulletin No. 1. Indian Institute of Soil Science, Bhopal, Madhya Pradesh.

Takkar, P.N., Singh, M.V. and Chaudhary, A.N. 1997. A critical review of plant nutrient supply needs, efficiency and policy issues of Indian agriculture for the year 2000: Micronutrients and trace elements. (In) *Plant Nutrient Needs, Supply, Efficiency and Policy Issues: 2000–2005.* Kanwar, J.S. and Katyal, J.C. (Eds). National Academy of Agricultural Sciences, New Delhi. pp. 238–264.

Velmurugan, R., Mahendran, P.P., Wani S.P., Uttam, K. and Prabhavathi, M. 2013. Molybdenum status and critical limit in the soil for green gram (*Vigna radiata*) growing in Madurai and Sivagangai districts of Tamil Nadu, India. *Soil Science and Plant Nutrition* **59**: 2,229–2,236.

Von Uexkull, H.R. and Sanders, J.J. 1986. Chlorine in the Nutrition of Palm Trees. (In) *Special Bulletin on Chloride and Crop Production* No. 2. Jackson T.L. (Ed.), Phosphate and Potash Institute, Atlanta, GA, USA.

Yadav, G..S., Shivay, Y.S., Kumar, D. and Babu, S. 2013. Enhancing iron density and uptake in grain and straw of aerobic rice through mulching and rhizo-foliar fertilization of iron. *African Journal of Agricultural Research* **8**(xx): 5,447–5,454.

Yakumin, A.F., Chan Van, N. and Gogitov, I.N. 1991. Effect of molybdenum, vanadium and tungsten on the growth of *Anabaena variabilis* and its synthesis of nitrogenases. *Microbiology* (New York) **60**: 52–56.

QUESTIONS

1. Fill in the blanks

(a) Total number of micronutrients essential for plant growth are ....
(b) Micronutrients absorbed as cations are (write only symbols): 1 ...... 2 ...... 3 ...... 4 ...... 5 ......
(c) Micronutrients absorbed as anions are (write only symbols): 1...... 2 ......
(d) Micronutrient absorbed in molecular form is ....
(e) Chlorine deficiency may be seen in ....
(f) Toxicity of Mn is reported in ....

2. Answer True (T) or False (F)

(a) Cu deficiency is a serious problem in Tamil Nadu. ....
(b) Zn deficiency is wide-spread in India. ....
(c) Legume plants absorb Fe with the help of Fe-Siderophores. ....
(d) Sulphur deficiency in plants is expressed in top leaves. ....
(e) Extreme Fe deficiency can result in bleached leaves. ....

Chapter 12

Zinc

Internationally maize (*Zea mays*) and citrus (*Citrus* sp.) are reported to largely suffer from Zn deficiency, for which these are considered as marker plants. Zinc deficiency in maize results in pale or whitish top young leaves and the deficiency is referred to as '*white bud*' (Camberato and Maloney, 2012). In citrus, Zn deficiency results in '*rosette*' (cluster of stiff leaves at the end of young shoots) formation (Brennan, 1993). Zinc deficiency in India was first reported in rice by Nene (1966) at the Govind Ballabh Pant University of Agriculture and Technology, Pantnagar, Uttarakhand, and given the name '*khaira*' disease due to dark-brown-coloured spots resembling *kattha* (*Acacia catechu* extract used in *paan* – the betel leaves) on the leaves of zinc deficient plants. Soon All India Coordinated Project on Micronutrients was launched by the Indian Council of Agricultural Research (ICAR) with headquarters at the Punjab Agricultural University, Ludhiana, which was later shifted to the Indian Institute of Soil Science, Bhopal, Madhya Pradesh. About half of the cultivated soils in the India are deficient in available zinc. Zinc deficiency is most in Maharashtra and Karnataka and rice-growing states of north and northeast (Table 12.1). A good response of cereals to Zn fertilization has been reported (Rattan *et al.*, 2008).

Table 12.1: Critical limits of DTPA-extractable zinc (mg kg^{-1} soil) for some crops in some soils

Soil	Rice	Wheat	Maize	Sorghum
Alluvial and Black	0.38–0.90	0.40–0.80	0.54–1.0 and 0.84–1.34	0.54
Red and Black Red	0.45–2.0	0.46–0.80	1.0–2.0	1.0–2.0
	0.60–1.0	–	0.65–0.80	–

Based on 15,000 trials on research stations, Singh (2008) reported the average response of different groups of crops as follows:

Cereals 420–550 kg grain ha^{-1} (15.7–23% increase over no-zinc control)

Pulses 170–460 kg grain ha^{-1} (7.3–28.2% increase over no-zinc control)

Oilseeds 110–360 kg grain ha^{-1} (11.4–40% increase over no-zinc control)

Fodders 0.09–4.6 tonnes green fodder ha^{-1} (5–34% increase over no-zinc control)

It may be added that crop varieties/ hybrids also differ in their response to Zn. For example, Chaudhary *et al.* (1977) reported that, 'IR-6' suffered due to Zn deficiency more than '*Basmati*-370' and Hossain *et al.* (2011) from Bangladesh reported that BARI Hybrid Maize-3 and composite BARI Maize-6 were more Zn responsive (Zn efficient) than hybrids 'BARI Hybrid Maize-1, BARI Hybrid Maize Top 1 and Pacific 984 and composites 'Mohor', 'Barnali', 'Khoibhutta'.

Most research on Zn has focused on yield and Zn uptake by crops. However, Prasad *et al.* (2014) focused the attention on fertilizing cereals for biofortification of grains with Zn and Fe and a number of experiments have been conducted with this objective and have shown that Zn fertilization is the easiest and fastest way to achieve this (Shivay *et al.*, 2008a,b; 2014; 2015a,b).

TOTAL ZINC AND FORMS OF ZINC IN SOIL

Total Zn in world soils ranges from 10 to 300 mg kg^{-1}, with an average of 50 mg kg^{-1} (Kiekens, 1995). In general, light sandy soils contain 31–61 mg Zn kg^{-1}, while heavy soil 35–75 mg Zn kg^{-1} (Noulas *et al.*, 2018). Total Zn in Indian soils is reported to be 47 mg kg^{-1} in Entisols, 60 mg kg^{-1} in Inceptisols, 61 mg kg^{-1} in Aridisols, 63 mg kg^{-1} in Vertisols, 44 mg kg^{-1} in Alfisols, 43 mg kg^{-1} in Ultisols, 30 mg kg^{-1} in Mollisols and 72 mg kg^{-1} in Oxisols (Katyal and Sharma, 1991).

Most Zn is present as insoluble compounds and is unavailable to plants. Only exchangeable Zn on clay minerals, hydrous oxides of Al and Fe and organic matter particle surfaces contributes to soil solution Zn and becomes available to crop plants. Exchangeable Zn in Indian soils varies from 0.1 to 2 mg kg^{-1} (Singh, 2011). The Zn in soil solution at any time is very little and tends to be in equilibrium with exchangeable Zn. Barber (1995) reported that, Zn in soil solution varies from 10^{-10} M to 10^{-6} M. Caroll and Loneragan (1969) reported that 10^{-7} to 10^{-9} M Zn levels in solution culture studies were found to be critical for plant growth. Deficiency of Zn is likely to occur in soils, when soil solution Zn is between 10^{-10} and 10^{-8} M (Lindsay, 1991).

SOIL TEST FOR AVAILABLE ZINC

In India, 0.005 M DTPA + 0.1 M TEA + 0.01 M $CaCl_2$ buffered *p*H 7.3 is the method used for determining available Zn in soils (Lindsay and Norvell, 1978) and the critical limits for some crops on different soils are given in Table 12.1. Deficiency of Zn is widespread all over the country (Fig.12.1), and has received considerable attention in rice–wheat cropping system (Prasad, 2005).

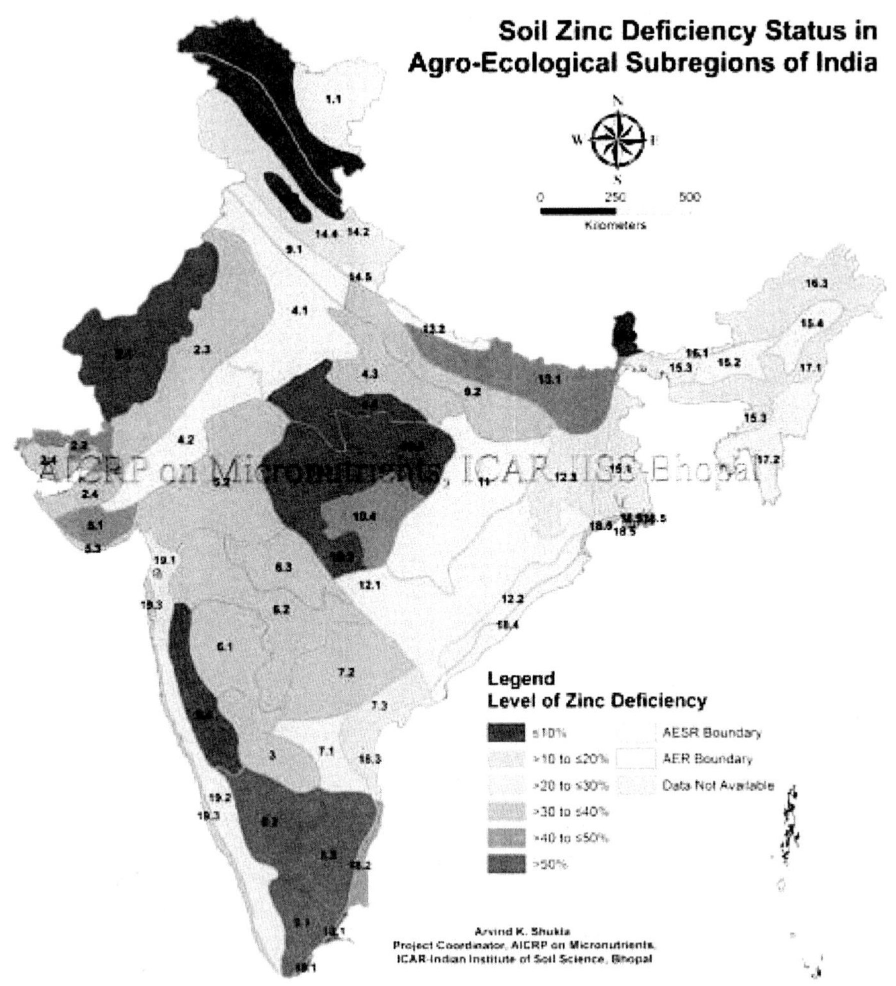

Fig. 12.1: Zn deficiency in different agro-ecological regions of India (with permission from the Project Coordinator, AICRP on Micronutrients, Indian Institute of Soil Science, Bhopal)

FACTORS AFFECTING ZINC AVAILABILITY

A number of factors affect Zn availability in soils. These include *p*H, presence of calcium carbonate, soil moisture, amounts of Na, Mg and P in soil solution, loss of soil organic matter, and plant root volume.

Soil *p*H

The solubility of Zn decreases 100-fold for each unit increase of soil *p*H (Lindsay, 1991). Rupa and Tomar (1999) from laboratory studies reported that, desorption of adsorbed P decreased with increasing *p*H and abruptly stopped at *p*H 7.5.

Calcium carbonate

Calcium carbonates increases soil *p*H, sorbs Zn on its surface and may form insoluble calcium zincate. Papadopoulos and Rowell (1989) reported that, Zn could be held on calcium carbonate surfaces as $Zn_5(OH)_6(CO_3)_2$. This would explain why Zn deficiency is more prevalent on calcareous soils.

Amounts of sodium, calcium, magnesium

Both Na and Mg increase soil *p*H and also compete with Ca for absorption sites on plant roots. Katyal and Sharma (1991) reported reduced Zn uptake by plants in Ca- and Mg-rich soils.

Phosphate

Zinc deficiency in corn is often aggravated by P fertilization, especially on calcareous soils (Langin *et al.,* 1962). Zinc can form insoluble zinc phosphate when concentration of phosphate is high in soil solution (Lambert *et al.*, 2007); however, this may still be available to crop plants. In field studies in India, zincated superphosphate compared well with zinc sulphate as a source of Zn (Sakal *et al.*, 1985). Customized zincated NPKs are produced and used in India.

Marschner (1993) observed that, root colonization by VA mycorrhizae increases spatial availability of zinc similar to that of phosphorus. Mycorrhizal plants usually have higher zinc contents in the shoot dry matter and are less sensitive to zinc deficiency than non-mycorrhizal plants. As a rule, all factors which impair root colonization by VA mycorrhizae, including high levels of soil or fertilizer phosphorus, tend to decrease zinc contents in plants and increase the risk of zinc deficiency in plants grown on soils low in extractable zinc.

Soil moisture

Adequate moisture in soil is necessary for Zn uptake. This is why Zn deficiency is more pronounced in dry regions. Gao *et al.* (2006) reported that, Zn concentration in rice leaves at tillering as well as physiological maturity was lower under alternate wetting and drying conditions as compared to that under continuous flooding.

Root system

Since Fe, Zn, Mn, Cu, and Ni all have restricted mobility in the soil, their acquisition can be improved by increasing root biomass and by developing a more extensive root-system (White and Greenwood, 2013). Roots play a greater part in the translocation of Zn from soil to grain during grain-formation stage in rice as compared to wheat and barley, because there is no discontinuation of xylem at the base of the grain in rice as it is in wheat and barley (Prasad and Shivay, 2018). As a contrast, most Zn in the grains in wheat and barley comes by translocation from the leaves.

ZINC MANAGEMENT

Sources of Zinc

The major sources of Zn in India are zinc sulphate heptahydrate ($ZnSO_4.7H_2O$) containing 21% Zn and zinc sulphate monohydrate ($ZnSO_4.H_2O$) containing 33% Zn. Though Zn-EDTA (12% Zn) is available, it is more expensive. Frits of Zn (10–30% Zn) can also be used. Shivay *et al.* (2015 a, b) have suggested the use of 0.5% zinc-coated urea as a source of Zn for foliar application of Zn, but it is not available in the market.

Rate of application

Most recommendations are made on the basis of soil-test values for available Zn. The general recommendations are 10–25 kg zinc sulphate heptahydrate (2–5 kg Zn ha^{-1}) or its equivalent.

Method and Time of Application

Most recommendations are for deep placement of the entire dose of zinc at sowing along with basal N, P and K. However, Shivay *et al.* (2015a) have reported that in addition to basal Zn application, one foliar spray of 0.5% zinc sulphate heptahydrate solution at flowering in rice increased rice yield, Zn uptake by rice crop and Zn concentration in rice grain. Thus, foliar application of Zn in addition to basal Zn application is important from the viewpoint of Zn biofortification of grains. For foliar spray, Zn-EDTA is a better choice (Shivay *et al.,* 2015b; Shivay *et al.*, 2016).

INTERACTION WITH OTHER NUTRIENTS

Zinc interacts positively with N and K and negatively with P, Ca, Fe and Cu (Prasad *et al.*, 2016). While P reacts with Zn to form less-soluble compound, Ca, Fe and Cu compete for the absorption sites on plant roots. It also interacts negatively with B, and Zn fertilization is recommended to overcome B toxicity (Gunes *et al.,* 2000).

SUMMARY

1. Total Zn in Indian soils is reported to be 30–72 mg kg^{-1} soil.
2. In India, 0.005 M DTPA + 0.1 M TEA + 0.01 M $CaCl_2$ buffered at *p*H 7.3 is the accepted extractant for determining available Zn in soils (Lindsay and Norvell method).
3. Factors affecting Zn availability in soil include: *p*H, Ca carbonate content, phosphorus content, Ca and Mg content, soil moisture and root structure.
4. The major sources of Zn in India is zinc sulphate heptahydrate ($ZnSO_4.7H_2O$), containing 21% Zn
5. Zinc is mostly applied to soil at sowing. If necessary, it can be applied to foliage, preferably as Zn–EDTA.
6. Zinc interacts positively with N and K.
7. Zinc interacts negatively with P, Ca, Fe Cu and B.

REFERENCES

Barber, S.A. 1995. *Soil Nutrient Bio-availability*, edn. 2, Wiley, New York.

Brennan, R.F., Armour, J.D. and Reuter, D.J. 1993. Diagnosis of zinc deficiency. (In) *Zinc in Soils and Plants,* pp. 167–183. Robson A.D. (Ed.), Kluwer, Dordrecht, Netherlands.

Camberato, J. and Maloney, S. 2012. Zinc deficiency in corn. Department of Agronomy, Purdue University, Soil Fertility Update, 4 p.

Carroll, M.D. and Loneragan, J.F. 1969. Response of plant species to concentration of Zn in solution. II. Rate of Zn absorption and their relation to growth. *Australian Journal of Agricultural Research* **20**: 457–463.

Chaudhary, F.M., Alam, S.M., Rashid, A. and Latif, A. 1977. Mechanism of differential susceptibility of two rice varieties to zinc deficiency. *Plant and Soil* **46**(3): 637–642.

Gao, X., Zou, C., Fan, X., Zhang, F. and Hoffland, E. 2006. From flooded to aerobic conditions in rice cultivation: consequences on zinc uptake. *Plant and Soil* **280**: 41–47.

Gunes, A., Alpaslam, M., Cikli, Y. and Ozen, H. 2000. The effect of zinc on alleviating boron toxicity. *Turkish Journal of Agriculture and Forestry* **24**: 506–509.

Hossain, M.A., Jahiruddin, M. and Khatun, F. 2011. Response of maize varieties to zinc fertilization. *Bangladesh Journal of Agricultural Research* **36**(3): 437–447.

Katyal, J.C. and Sharma, B.D. 1991. DTPA extractable and total Zn, Cu, Fe and Mn in Indian soils. *Geoderma* **49**: 165–179.

Kiekens, L. 1995. Zinc. (In) *Heavy Metals in Soils*, edn 2, pp. 284–305. Allway, B.J. (Ed.), Blackie Academic & Professional, London.

Lambert, R., Grant, C. and Sauve, S. 2007. Cadmium and zinc in soil solution extracts following the application of phosphate fertilizers. *Science of the Total Environment* **378**: 298–305.

Langin, W.J., Ward, R.C., Olson, R.A. and Rhoades, H.F. 1962. Factors responsible for poor response of corn and grain sorghum to phosphorus fertilization: II. Lime and P placement effects on P–Zn relations. *Soil Science Society of America Proceedings* **26**: 574–578.

Lindsay, W.I. 1991. Inorganic equilibria affecting micronutrients in soil. (In) *Micronutrients in Agriculture.* Mortvedt J.J., Fox F.R., Schuman L.M. and Welch R.M., (Eds). Soil Science Society of America, Madison, WI, the USA.

Lindsay, W.I. and Norvell, W.A. 1978. Development of DTPA soil test for zinc, iron, manganese and copper. *Soil Science Society of America Journal* **42**: 421–448.

Marschner H.1993. Zinc uptake from soils. (In) *Zinc in Soils and Plants: Developments in Plant and Soil Sciences,* vol. 55, pp. 59–77. Robson A.D. (Ed.), Springer, Dordrecht.

Nene, Y.L. 1966. Symptoms, causes and control of *khaira* disease of paddy. *Bulletin of the Indian Phytopathology Society* **3**: 97–101.

Noulas, C., Tziouvalekas, M. and Karyotis, T. 2018. Zinc in soils, water and food crops. *Journal of Trace Elements in Medicine and Biology* **49**: 252–280.

Papadopoulos, A. and Rowell, D. 1989. The reactions of copper and zinc with calcium carbonate surfaces. *Journal of Soil Science* **40**(1): 39–48.

Prasad, R. 2005. Rice–wheat cropping systems. *Advances in Agronomy* **86**: 255–339.

Prasad, R. and Shivay, Y.S. 2018. Relative grain zinc loading ability of cereals (rice, wheat, maize) and a grain legume (chickpea). *Indian Journal of Agricultural Sciences* **88**(2): 307–313.

Prasad, R., Shivay, Y.S. and Kumar, D. 2014. Agronomic biofortification of cereal grains with iron and zinc. *Advances in Agronomy* **125**: 55–91.

Prasad, R., Shivay, Y.S. and Kumar, D. 2016. Interactions of zinc with other nutrients in soils and plants–A review. *Indian Journal of Fertilisers* **12**(5): 16–26.

Rattan, R.K., Datta, S.P. and Katyal, J.C. 2008. Micronutrient management–research achievements and future challenges. *Indian Journal of Fertilisers* **4**(12): 93–118.

Rupa, T.R. and Tomar, K.P. 1999. Zinc desorption kinetics as influenced by *p*H and phosphorus in soils. *Communications in Soil Science and Plant Analysis* **30**: 1,951–1,962.

Sakal, R.A., Singh, A.P. and Singh, B.P. 1985. Use of zincated superphosphate to control Zn deficiency in calcareous soils. *Journal of the Indian Society of Soil Science* **33**: 443–446.

Shivay, Y.S., Kumar, D. and Prasad, R. 2008a. Effect of zinc enriched urea on productivity, zinc uptake and efficiency of aromatic rice–wheat cropping system. *Nutrient Cycling in Agroecosystems* **81**: 229–243.

Shivay, Y.S., Prasad, R., Kaur, R. and Pal, M. 2016. Relative efficiency of zinc sulphate and chelated zinc on zinc bio fortification of rice grains and zinc use-efficiency in *Basmati* rice. *Proceedings of the National Academy of Sciences, India, Section B Biological Sciences* **86**(4): 973–984.

Shivay, Y.S., Prasad, R and Pal, M. 2014. Effect of conditioning zinc sulphate hepta hydrate (ZnSHH) and zinc oxide (ZnO) with neem oil on growth, productivity, zinc biofortification and zinc uptake by *Basmati* rice. *Journal of Plant Nutrition* **37**: 1,873–1,874.

Shivay, Y.S., Prasad, R. and Pal, M. 2015b. Effect of source and method of zinc application on yield, zinc biofortification of grain, and Zn uptake and use efficiency in chickpea (*Cicer arietinum* L). *Communications in Soil and Plant Analysis* **46:** 2,191–2,200.

Shivay, Y.S., Prasad, R. and Rahal, A. 2008b. Relative efficiency of zinc oxide and zinc sulphate coated urea for spring wheat. *Nutrient Cycling in Agroecosystems* **82**: 259–264.

Shivay, Y.S., Prasad, R., Singh, R.K. and Pal, M. 2015a. Relative efficiency of zinc-coated urea and soil and foliar application of zinc sulphate on yield, nitrogen, phosphorus, potassium, zinc and iron biofortification in grains and uptake by *Basmati* rice (*Oryza sativa* L.). *Journal of Agricultural Science,* Canada **7**(2): 161–173.

Singh, M.V. 2008. Micronutrient deficiencies in soils and crops of India. (In) *Micronutrient Deficiencies in Global Crop Production,* Alloway J.B. (Ed.). Springer Sciences Business Media.

Singh, M.V. 2011. Assessing zinc deficiency for soil factors affecting and nutritional scarcity in humans and animals. *Indian Journal of Fertilisers* **7**(10): 36–43.

White, P.J. and Greenwood, D.J. 2013. Properties and management of cationic elements for crop growth. (In) *Soil Conditions and Plant Growth,* pp. 160–194. Gregory, P.J. and Nortcliff, S. (Eds), Blackwell Publishing, Oxford, UK.

QUESTIONS

1. Fill up the blanks

(a) The critical level of Zn in soil solution is ....
(b) The accepted extractant for determining available Zn in soils ....
(c) The nutrient (other than P) most affecting Zn availability in soil is ....
(d) Most common Zn fertilizer is ...., which contains .... % Zn.
(e) The most important soil factor affecting the availability of Zn is ....

2. Answer true (T) or false (F)

(a) *White bud* in maize and *khaira* disease in rice are caused by Zn deficiency.
(b) Foliar application is the best method of Zn fertilization.
(c) Zn application does not increase Zn content in rice.
(d) Due to higher leaching losses, rates of Zn application are higher on sandy soils.
(e) Zinc deficiency in humans is not of much importance.

Chapter 13

Boron

Boron (B) deficiency has been realized as the second most important micronutrient constraint in crop production after that of Zn in the world (Ahmed *et al.*, 2012). However, in Nepal, B deficiency is more wide-spread than Zn deficiency, as 80 to 90% of soil samples were deficient in B, 20 to 50% in Zn and 10 to 15% in Mo (Anderson, 2007). In India, B deficiency has increased from 2% of the soil samples deficient in B in 1980s (Katyal and Vlek, 1995) to 52% soil samples in 2000s (Singh, 2012). Total B content in Indian soils varies from 7 to 630 mg kg^{-1} soil (Kanwar, 1976); being the highest in arid regions of the country and the lowest in wet humid regions of the country (Behera *et al.*, 2009; Prasad *et al.*, 2014). Boron deficiency is most in the heavy rainfall areas of Western *Ghats* of Maharashtra, Karnataka and the eastern states of Bihar, Odisha, Jharkhand and Chhattisgarh states of India (Fig.13.1). The wide-spread deficiency of B in India can be seen from the fact that B fertilization recorded the highest benefit : cost (B : C) ratio in rice–wheat cropping system at Palampur (Himachal Pradesh), Ludhiana (Punjab), Faizabad, Modipuram and Varanasi (Uttar Pradesh), Pantnagar (Uttarakhand) and in rice–rice cropping system at Maruteru (Andhra Pradesh), Karjat (Maharashtra) and Jorhat (Asom) (Singh and Goswami, 2014). Response of wheat to B fertilization is also reported from Bangladesh (Halder *et al*, 2007).

Boron is a specialty micronutrient, because it is related with fruit quality in many fruit and vegetable crops, such as guava– *Psidium guajava* L. (Bhatia *et al.*, 2001; Awasthi and Lal, 2009), papaya–*Carica papaya* L. (Mollah *et al.*, 2006); date palm–*Phoenix daclylifera* L. (Khayyat *et al.*, 2007), litchi–*Litchi chinensis* sonner (Alila and Achumi, 2012), apple–*Malus pumila* Mill.; syn *M. domestica* Borkh. (Wojcik *et al.*, 1999) and tomato–*Solanum lycopersicum* L. (Davis *et al.*, 2003). Eaton (1932) reported that among the field crops, cotton required the highest amount of B and was very sensitive to B deficiency leading to abscission and shedding of squares and bolls. Lancaster *et al.* (1962) recommended B application to cotton in Mississippi (USA). Boron deficiency has also been reported as a cause of abscission and shedding of squares and bolls in cotton in India (Fig 13.2) (Perumal *et al.*, 2006). Singh and Goswami (2014) also reported abscission of flower and reduced grains per panicle in rice due to B deficiency. Kataki *et al.* (2001) from Nepal reported that, soil application of B at sowing reduced the sterility in wheat by more than 50% and doubled the yield by increasing seed set.

Other than cotton, the groups of crops are in the following order in respect of B uptake: oilseeds > pulses > tubers > cereals. This pattern is somewhat similar to S uptake by crops, both being taken up as anions. Some data are in Table 13.1.

Table 13.1: Boron uptake (g t^{-1} grain/seed/tubers)

Group-Crop	B uptake
Oilseeds	
Rapeseed-mustard	138–214
Sunflower	125
Groundnut	43
Castor	19
Sesame	170–347
Cereals	
Rice	11–16
Wheat	31–48
Sorghum	54
Pearl millet	56
Maize	50–134
Pulses	
Chickpea	57–132
Black gram	189
Pigeonpea	105
Green gram	52
Lentil	62–80
Tubers	
Potato	50
Sweet potato	33
Others	
Onion	20–22
Tea	101

Source: Singh and Goswami (2014)'

PRIMARY BORON MINERALS

Boron (B) in soils comes from primary B minerals. The primary B minerals are: tourmaline, ulexite [$NaCa(B_5O_6)(OH)_6.5H_2O$], borax [$Na_2B_4O_5(OH)_4.8H_2O$], colemanite [$Ca(B_3O_4.(OH)_3.2H_2O$ and kernite [$Na_2B_4O_5(OH)_3.2H_2O$]. The tourmaline mineral group is chemically one of the most complicated groups of silicate minerals. Its composition varies widely because of great isomorphous substitution. Its general formula is:

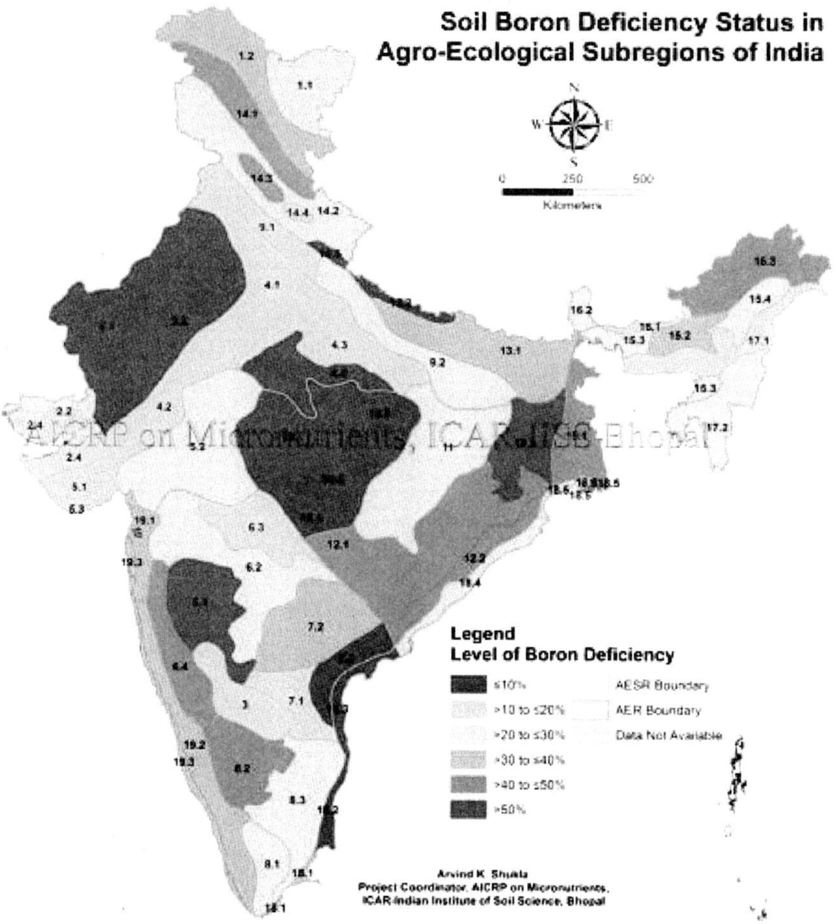

Fig. 13.1: Extent of B deficiency in different agro-ecological zones of India.

(With permission from the Project Coordinator, AICRP on Micronutrients, Indian Institute of Soil Science, Bhopal)

Fig. 13.2: Early stage of abscission of cotton square. Photo courtesy ICAR-Central Institute of Cotton Research, Nagpur, Maharashtra with permission (Perumal *et al.*, 2006)

PRIMARY BORON MINERALS

Boron (B) in soils comes from primary B minerals. The primary B minerals are: tourmaline, ulexite [$NaCa(B_5O_6)(OH)_6.5H_2O$], borax [$Na_2B_4O_5(OH)_4.8H_2O$], colemanite [$Ca(B_3O_4.(OH)_3.2H_2O$ and kernite [$Na_2B_4O_5(OH)_3.2H_2O$]. The tourmaline mineral group is chemically one of the most complicated groups of silicate minerals. Its composition varies widely because of great isomorphous substitution. Its general formula is:

$$XY_3Z_6(T_6O_{18})(BO_3)_3V_3W$$

where **X** = Ca, Na, K; **Y** = Li, Mg, Fe^{2+}, Mn^{2+}, Zn, Al, Cr^{3+}, V^{3+}, Fe^{3+}, Ti^{4+}; **Z** = Mg, Al, Fe^{3+}, Cr^{3+}, V^{3+}; **T** = Si, Al, B; **V** = OH, O and **W** = OH, F, O (Hawthorne and Henry, 1999). The name tourmaline comes from the Sinhalese word "*thoramalli*". India does not have its own deposits and most raw ores of B minerals and B fertilizers are imported, mostly from the USA.

FORMS IN SOIL

Most B in soil is attached to SOM, which is a major contributor. As regards inorganic forms, B may be present as: (*i*) primary minerals; (*ii*) in the exchange complex of clay minerals; (*iii*) sesquioxide bound; and (*iv*) small amounts in soil solution. In 20 slightly alkaline soils of groundnut-growing areas, the mean concentration of B in soluble, adsorbed, organic bound and sesquioxide bound B was 0.3, 0.11, 0.21 and 0.24% of total B in soil (Singh and Goswami, 2014). Contribution of different fractions to crop plants will vary from soil to soil and may also vary from crop to crop on the same soil.

In soil solution, B is present as boric acid, which is a weak (pKa=8.98) Lewis acid (acids that can accept a proton). H_3BO_3 is acidic because of its reaction with OH^- from water, forming the tetrahydroxyborate complex [$B(OH)_4$] and releasing the corresponding proton left by the water autoprotolysis as shown below:

$$H_3BO_3 + 2H_2O \longrightarrow [B(OH)_4]^- + [H_3O]^+$$

Undissociated H_3BO_3 is the form in which it is taken up by crop plants (Binham *et al.* 1970). As mass flow and based on solubility criteria its concentration in soil solution is $10^{-5.5}$ M B (Lindsay, 1991). It is the tetrahydroxyborate complex [$B(OH)_4$] form in which it gets attached to the clay minerals and is the highest at *p*H 9.0 close to its *p*Ka value of 8.98 (Hingston, 1972). This would explain why liming acid soil leads to B fixation by clays.

SOIL TESTS FOR AVAILABLE BORON

A number of soil tests for available boron including extraction with hot water (Berger and Truog, 1939), 1 N ammonium acetate (Gupta and Stewart, 1975), 0.1 N calcium chloride (Gupta and Stewart, 1975), 0.1 M calcium chloride + 0.01 M Mannitol (Cartwright *et al.*, 1983). In India, hot-water extraction method is used for determining available B and is abbreviated as HWS-B. Although a number of researchers have reported different critical deficiency levels (CDL) for available B in soils, 0.5 mg kg^{-1} soil HWS-B is the accepted value for declaring soils deficient in B (Singh and Goswami, 2014). Based on this critical value, the percentage of soil samples testing low HWS-B in different states are given in Table 11.1; in general, higher values were recorded for the states having leached acid red and lateritic soils. However, Singh (2007) suggested categorizing HWS-B into following 5 classes for making recommendations for B fertilization. These are:

< 0.25 mg kg^{-1} soil HWS-B: 125% of the optimum dose recommended for a crop in the area.

0.25–0.50 mg kg^{-1} soil HWS-B: 100% of the optimum dose recommended for a crop in the area.

The 0.50–0.75 mg kg^{-1} soil HWS-B: 50% of the optimum dose recommended for a crop in the area.

The 0.75–1.00 mg kg^{-1} soil HWS-B: 25% of the optimum dose recommended for a crop in the area.

The > 1.00 mg kg^{-1} soil HWS-B: No need for B fertilization.

BORON TOXICITY

Phyto-toxicity due to B can occur, when HWS-B in soils exceeds 5 mg kg^{-1} soil, and soils with 100 mg kg^{-1} soil HWS-B or more are considered B toxic. Boron-toxicity symptoms include leaf necrosis and reduced leaf area.

Boron content in irrigation water also needs to be kept in view. Safe irrigation water to avoid B toxicity and yield loss was reported to have less than 1.0 mg B L^{-1} for sunflower (*Helianthus annuus* L.), 3–4 mg B L^{-1} for wheat and 4–6 mg B L^{-1} for peas on a sandy-loam soil (Chauhan and Powar, 1978).

CROP SENSITIVITY TO BORON DEFICIENCY AND TOXICITY

The concentration range between deficiency and toxicity of B is quite narrow and special care is required while recommending B fertilization. Critical toxicity limit (mg kg^{-1} dry matter) for some crops are: maize 100, wheat 100–270, cucumber (*Cucumus sativus* L.) 400, squash (*Cucurbita pepo* L.)

1000 (Romheld and Marschner, 1991). On the basis of susceptibility to B deficiency, the crops are grouped into following 3 classes (Gupta, 2005; Singh and Goswami, 2014).

Less susceptible

Cereals, viz. wheat, tall local rice varieties, oats, barley, sorghum [*Sorghum bicolor* (L.) Moench], pearl millet [*Cenchrus americanus* (L.) Morrone], finger millet [*Eleusine coracana* (L.) Gaertn.]; pulses and oilseeds, viz. pea, pigeonpea [*Cajanus cajan* (L.) Millsp.], peas, beans, soybean); plantation crops, viz. pineapple [*Ananas comesus* (L.) Merr.], rubber [*Hevea brasilicensis* (Willd: ex Juss.) Mull. Arg].

Moderately susceptible

Cereals, viz. maize, high yielding varieties and hybrid rice; pulses and oilseeds, viz. blackgram, greengram, mustard, groundnut; Vegetables, viz. radish (*Raphanus sativus* L.), onion, cabbage, tomato, potato; cash crops, viz. tobacco (*Nicotiana tabaccum* L.), sugarcane; fruit crops, viz. pomegranate (*Punica granatum* L.), citrus, peach [*Prunus persica* (L.) Batsch], pear (*Pyrus* sp.), banana (*Musa paradisiaca*), papaya (*Carica papaya* L.), coconut (*Cocos nucifera* L.); plantation crop, viz. tea [*Camellia sinesis* (L.) Kuntze].

Highly susceptible

Cash crop, viz. cotton; pulse, viz. chickpea; oilseeds, viz. rapeseed (*Brassica napus* L.), sunflower; vegetables, viz. cauliflower, carrot [*Daucus carota* sub. sp. *sativus* (Hoffin.) Schubl. C. Martins], turnip (*Brassica sapa* subsp. sapa), sugar beet; forages, viz. alfalfa (*Medicago sativa* L.), berseem (*Trifolium alexandrinum* L.), fruit crops, viz. apple, grapes (*Vitis vinifera* L.), olive (*Olea europaea* L.), plantation crop, viz. coffee (*Coffea arabica* L.), flower crops, viz. marigold (*Tagetes erecta* L.), chrysanthemum.

BORON MANAGEMENT

Sources

Borax [$Na_2B_4O_7.10H_2O$] containing 11.5% B and boric acid [H_3BO_3] containing 17% B used to be the sources of B in India, but gradually new generation B-fertilizers, such as, crystalline Fertibor® [$Na_2B_4O_7.5H_2O$] and granular Garnubor® [$Na_2B_4O_7.5H_2O$] each containing 14.6% B and solubor® [$Na_2B_8O_{13}.4H_2O$] containing 20% B are becoming more popular, owing to their reduced transport costs on account of higher B content. Borax, Fertibor® and Garnubor® are all disodium borate and differ only in water molecule content, which affects its B content. A list of B fertilizers marketed in India is given in Table 13.2.

Rate of boron application

Boron recommendations are made on the basis of soil-test values as discussed earlier. The general recommendations on B-deficient soils for irrigated crops are: 0.5 kg B ha^{-1} for wheat, greengram, blackgram, chickpea, lentil (*Lens culinaris* Medik.), soybean; 0.75 kg B ha^{-1} for pigeonpea, rapeseed/ mustard, potato, sesame (*Sesamum indicum* L.), groundnut; 1.0 kg B ha^{-1} for rice, maize, cotton, castor (*Ricinus communis* L.), sugarcane, cauliflower (*Brassica oleracea* var. *botrytis* L.), cabbage (*Brassica oleracea* var. *capitata* L.) (Singh and Goswami, 2014).

Table 13.2: Boron fertilizers marketed in India

B–fertilizers	B (%)	Other nutrients (%) [$N + P_2O_5 + K_2O$]	Dose (kg ha^{-1})	Application mode
Straight B fertilizers				
Garnubor®	14.6	–	3–7	Soil – at planting
Fertibor®	14.6	–	3–7	Soil – at planting/ foliar
Borax	10.5	–	7–10	Soil – at planting/ foliar
Colemanite	11	–	10–15	Soil – acid soils only
Solubor®	20	–	1–2 g L^{-1}	Foliar
Boric acid	17	–	1–2 g L^{-1}	Foliar
Coated or fortified				
Boronated SSP	0.15–0.20	16 + 0 + 0	As per P dose	Soil – at planting
Boronated DAP	0.1–0.3	18 + 46 + 0	As per P dose	Soil – at planting/top dressing
Boronated NPKs	0.2	15 + 15 + 15	As per P dose	Soil – at planting/top dressing
	0.3	10 + 26 + 26		
	0.3	12 + 32 + 16		
	0.2	24 + 24 + 0		

Source : Singh and Goswami (2014)

Time and Method of Application

Application of the entire recommended dose at planting is the best practice. If the basal application is missed, top-dressing can be done up to 45 days after planting. Foliar application is not as effective as the soil application. However, when necessary, 2 to 3 sprays at a gap of 2–3 weeks, starting at 45–60 days after germination/ transplanting and later at pre-flowering of 0.15–0.20% solution of Solubor® or boric acid is recommended (Singh and Goswami, 2014).

INTERACTIONS WITH OTHER NUTRIENTS

Boron interacts positively with N and K and negatively with Zn. Some data on N × B interaction are in Table 13.3. Data in Table 13.3 show that when no N was applied, application of 1.12 kg B ha^{-1} increased the number of bolls $plant^{-1}$ by 1.0, while in the presence of 80 or 160 kg N ha^{-1} the increase was 3.4.

Table 13.3: Bolls $plant^{-1}$ in cotton as affected by nitrogen × boron interaction

B (kg ha^{-1})	0 kg N ha^{-1}	80 kg N ha^{-1}	160 kg N ha^{-1}
0	14.4^{a}	16.2^{c}	17.3^{c}
0.56	14.8^{a}	17.3^{b}	18.4^{b}
1.12	15.4^{a}	19.6^{a}	20.7^{a}

Means followed by different superscripts within columns are significantly different (P =0.05)
Source: Gormuz (2005)

Table 13.4: Effect of zinc fertilization on boron concentration in shoots (mg g^{-1} tissue) of lemon seedlings

Zn µg g^{-1} soil	B 0 µg g^{-1} soil	B 20 µg g^{-1} soil
0	23.9^{c}	699^{a}
10	21.7^{c}	523^{b}

Means followed by different letters within columns are significantly different (P=0.05)
Source: Rajaie *et al.* (2009).

Negative interaction between Zn and B is well known. Some data on B concentration in lemon [*Citrus limon* (L.) Burm. fil.] seedlings as affected by B × Zn interaction are given in Table 13.4, which show that B concentration in lemon seedlings significantly declined with Zn application (Rajaie *et al.*, 2009). Singh *et al.* (1990) also reported that B application increased the tissue concentration and uptake of B by wheat more in the absence than in the presence of Zn fertilization. That is why Zn fertilization is recommended to overcome B toxicity (Gunes *et al.*, 2000).

Interestingly, the integrated application of secondary nutrient S and micronutrients Zn and B, increased the yield of maize, castor, groundnut and greengram by 75, 52, 53 and 65%, respectively, over the farmers' practices in red and black soils (Rego *et al.*, 2007). This again brings out the fact that all essential nutrients are complementary and interact positively, when applied in the amounts needed by a crop.

SUMMARY

1. Boron deficiency in the world as well as in India is second to Zn-deficiency.
2. 42% Indian soils are B deficient as compared to 48% being Zn-deficient.
3. Boron is involved in pollen-tube development and pollen formation.
4. Boron deficiency leads to flower and fruit shedding.
5. Boron deficiency leads to poor-quality fruits.
6. Organic C is a major contributor to B in plant nutrition.
7. Some B is held in the exchange complex of soil.
8. Available B in soil is determined by hot water extraction and the general critical deficiency limit for HWS-B is 0.5 mg kg^{-1} soil.
9. Solid straight B fertilizers are boric acid (17% B), borax (10.5% B), solubor® (20.5% B), Garnubor® (14.5% B), Fertibor® (14.5% B).
10. In addition to straight B fertilizers, Boronated-SSP, Boronated-DAP and a number of Boronated-NPKs are also available. They contain about 0.2–0.3% B. When these fertilizers are used, the rate of B application is determined by the rate of P application.
11. Rate of B application generally vary from 0.5 to 1.0 kg B ha^{-1}.
12. Boron should preferably be applied at planting; however, if missed it can be top-dressed within 40–45 days after transplanting/planting.
13. Foliar application of B is not as effective as soil application. However, if not applied earlier and deficiency is observed a 0.15–0.20% solution of boric acid or borax can be sprayed.
14. As regards B uptake, the crops are in the following order: oilseeds > pulses > cereals.
15. In general, cereals remove 10–50 g B t^{-1} grain, while pulses and oilseeds remove 50–200 g B t^{-1} grain or even more.
16. Boron interacts positively with N and negatively with Zn.
17. Major B minerals are tourmaline, ulexite [NaCa $(B_5O_6)(OH)_6.5H_2O$], borax [$Na_2B_4O_5$ $(OH)_4.8H_2O$], colemanite [$Ca(B_3O_4)(OH)_3.2H_2O$ and kernite [$Na_2B_4O_5(OH)_3.2H_2O$]. These are mentioned, because they are used as fertilizers.
18. While Na-B minerals are water-soluble, Ca-B mineral colemanite is not so water soluble and is therefore recommended only for acid soil.

REFERENCES

Ahmed, W., Zia, M.H., Malhi, S.S., Niaz, A. and Ullah, S. 2012. Boron deficiency in soils and crops: A review. (In) *Crop Plant.* Gupta A. (Ed.), INTECH Open Access Publisher, (https://www.intechopen.com/books/ crop-plant/boron-deficiency-in-soils-and-crops-a-review).

Alila, P. and Achumi, I. 2012. Pre-harvest chemical treatments affect post-harvest quality of litchi fruit. *Acta Horticulturae* **934**: 755–761.

Andersen, A. 2007. A review of micronutrient problems in the cultivated soils of Nepal. *Mountain Research and Development* **27**(4): 331–335.

Awasthi, P. and Lal, S. 2009. Effect of calcium, boron and zinc foliar sprays on the yield and quality of guava (*Psidium guajava* L.). *Pantnagar Journal of Research* **7**(2): 223–224.

Behra, S.K., Singh, M.V. and Lakaria, B.L. 2009. Micronutrients deficiencies in Indian soils and their amelioration through fertilization. *Indian Farming* **59**(2): 28–31.

Berger, K.C. and Truog, E. 1939. Extraction and determination of plant available boron in soils. *Soil Science* **193**: 149–168.

Bhatia, S.K., Yadav, S., Ahlawat, V.P. and Dahiya, S.S. 2001. Effect of foliar application of nutrients on the yield and fruit quality of winter season guava cv. L-49. *Haryana Journal of Horticultural Science* **30**(1/2): 6–7.

Bingham, F.T., Elseewi, A. and Oertli, J.J. 1970. Characteristics of boron adsorption by excised barley roots. *Soil Science Society of America Journal* **34**: 613–617.

Cartwright, B., Tiller, K.G., Zarcinaz, B.A. and Spouncer, L.R. 1983. The chemical assessment of the boron status of soils. *Australian Journal of Soil Research* **21**: 321–332.

Chauhan, R.P.S. and Powar, S.L. 1978. Tolerance of wheat and pea to boron in irrigation water. *Plant and Soil* **50**: 145–149.

Davis, J.M., Sanders, D.C., Nelson, P.V., Lengnick, L. and Sperry, M.J. 2003. Boron improves growth, yield, quality and nutrient content of tomato. *Journal of American Society of Horticulture Science* **128**(3): 441–446.

Eaton, F.M. 1932. Boron requirements of cotton. *Soil Science* **34**: 302–305.

Gormuz, O. 2005. Interactive effect of nitrogen and boron on cotton yield and fiber quality. *Turkish Journal of Agriculture and Forestry* **29**: 51–59.

Gunes, A., Alpaslam, M., Cikli, Y. and Ozen, H. 2000. The effect of zinc on alleviating boron toxicity. *Turkish Journal of Agriculture and Forestry* **24**: 506–509.

Gupta, U.C. 2005. *Boron and its Role in Crop Production*. CRC Press. Boca Raton, FL, the USA.

Gupta, U.C. and Stewart, J.W.B. 1975. The extraction and determination of plant available boron in soils. *Schweizerische Landwirtschaftliche Forschung* **14**: 153–169.

Halder, N.K., Hossain, M.A., Siddiky, M.A., Nasreen N. and Ullah, M.H. 2007. Response of wheat varieties to boron application in calcareous brown flood plain soil at southern region of Bangladesh. *Journal of Agronomy* **6**: 21–24.

Hawthorne, F.C. and Henry, D.J. 1999. Classification of the minerals of the tourmaline group. *European Journal of Mineralogy* **11**: 201–215.

Hingston, F.J., Posner, A.M. and Quirk, J.P. 1972. Anion adsorption by goethite and bibbsite. I. The role of proton in determining adsorption envelops. *Journal of Soil Science* **23**: 177–193.

Kanwar, J.S. 1976. Micronutrients. (In) *Soil Fertility: Theory and Practices*. Kanwar, J.S. (Ed.). Indian Council of Agricultural Research, New Delhi.

Kataki, P.K., Upreti H.K. and Bhatta, M.R. 2001. Soil boron deficiency induced wheat sterility in Nepal. *Journal of New Seeds* **3**(4): 23–39.

Katyal, J.C. and Vlek, P.L.G. 1995. Micronutrient problem in tropical Asia. *Fertilizer Research* **7**: 69–94.

Khayyat, M., Tafazoli, E., Eshghi, S. and Rajae, S. 2007. Effect of nitrogen, boron, potassium and zinc sprays on yield and fruit quality of date palm. *American-Eurasian Journal of Agricultural and Environmental Science* **2**(3): 289–296.

Lancaster, J.D., Murphy, B.C. and Hurt (Jr), B.C., Arnold, B.L., Coats, R.E., Albritton, R.C. and Walton, L. 1962. Boron now recommended for cotton. *Mississippi Agricultural Experiment Station Bulletin* 635.

Lindsay, W.I. 1991. Inorganic equilibria affecting micronutrients in soil. (In) *Micronutrients in Agriculture*, Mortvedt, J.J., Fox, F.R., Schuman, L.M. and Welch, R.M. (Eds). Soil Science Society ofAmerica, Madison, WI, the USA.

Mollah, M.R.A., Rahim, M.A., Islam, M.J., Khatun, M.R. and Rahman, M.M. 2006. Response of papaya varieties to basal and foliar application ofboron. *International Journal ofSustainable Agriculture and Technology* **2**(1): 28–31.

Perumal, N.K., Hebbar, K.B., Rao, M.R.K. and Singh, P. 2006. Physiological disorders in cotton. *CICR Technical Bulletin* 28, ICAR-Central Institute of Cotton Research, Nagpur.

Prasad, R., Kumar, D., Shivay, Y.S. and Rana, D.S. 2014. Boron in Indian agriculture–A review. *Indian Journal of Agronomy* **59**(4): 511–517.

Rajaie, M., Ejraie, A.K., Owailaie, H.R. and Tavakoli, A.R. 2009. Effect of zinc and boron interaction on growth and mineral composition of lemon seedlings in a calcareous soil. *International Journal of Plant Production* **3**(1): 1,735–8,043. ISSN: 1735–6814(Print), (online).

Rego, T.J., Sahrawat, K.L., Wani, S.P. and Pardhasardhi, G. 2007. Widespread deficiencies of sulphur, boron and zinc in Indian semi-arid tropical soils. *Journal of Plant Nutrition* **30**: 1,569–1,573.

Romheld, Y. and Marschner, H. 1991. Functions of micronutrients in plants. (In) *Micronutrients in Agriculture*, edn 2. Mortvedt J.J., Fox F.R., Schuman L.M. and Welch R.M. (Eds). Soil Science Society of America, Madison, WI, the USA.

Singh, J.P., Dahiya, D.J. and Narwal, R.P. 1990. Boron uptake and toxicity in wheat in relation to zinc supply. *Fertilizer Research* **24**(2): 105–110.

Singh, M.V. 2007. *Micro and Secondary and Pollutant Element Research in India.* AICRP on Micro and Secondary and Pollutant Elements in Soil and Plant, Indian Institute of Soil Science, Bhopal **31**: 1–112.

Singh, M.V. 2012. Boron deficiency in crops and its correction in India. (In) *Proceedings of the National Seminar on Micronutrient Zinc and Boron Management*, organized by OUAT, Bhubaneshwar–Rio Tinto India, at Bhubaneshwar, 27 July 2012.

Singh, M.V. and Goswami, V. 2014. Boron management. (In) *Textbook of Plant Nutrient Management,* pp. 188–213. Prasad R., Kumar D., Rana D.S., Shivay Y.S. and Tewatia R.K. (Eds.). Indian Society ofAgronomy, New Delhi.

Wojcik, P., Cieslinski, G. and Mika, A. 1999. Apple yield and fruit quality as influenced by boron applications. *Journal of Plant Nutrition* **22**(9): 1,365–1,377.

QUESTIONS

1. Answer True (T) or False (F)

(a) Boron deficiency is most on acid red and laterite soils.
(b) Irrigation water having a B content of 3–4 mg B L^{-1} is not safe for wheat.
(c) Borax contains 11% B.
(d) Rapeseed crop removes about 200 g B t^{-1} of seed produced.
(e) Boric acid is the most popular B fertilizer.
(f) Toumaline minerals are found in plenty in Sri Lanka.
(g) For foliar application solubor® is a good material.
(h) Fruit shedding could be due to B deficiency.
(i) Pulses have lesser B requirement than cereals.
(j) Cotton is highly susceptible to B deficiency.

2. Match the followings

(a)	Colemanite	(a)	0.5 mg kg^{-1} soil HWS-B.
(b)	Solubor®	(b)	Sugarbeet
(c)	Rate of B application to wheat	(c)	Foliar spray of B.
(d)	Critical deficiency level of B	(d)	Calcium borate
(e)	Boron tolerant crop	(e)	1 kg ha^{-1}

Chapter 14

Organic Manures

Before the discovery of chemical fertilizer in early 20th century, organic manures and growing of legumes were the ways of replenishing soil fertility. The only other way was leaving land fallow for a few years before its re-cultivation, a practice followed in shifting cultivation (*jhuming*) (Garbyal, 1999). This was also true for India, where although the first single superphosphate plant was set up in 1906 in Ranipet (Tamil Nadu), the real use of fertilizer started with the manufacture of urea, the first plant for which was set up in 1959 at Sindri (Bihar) (Tandon, 2014). Organic manures could meet crop-nutrient needs in early days because the average grain yield of most cereals was about 1.5–2.0 t ha^{-1}, while the average yield of pulses was 0.5–0.75 t ha^{-1}. However, the crop production scenario changed rapidly after the introduction of dwarf, high-yielding varieties of wheat from Mexico, which led to the "Green Revolution" (Swaminathan, 2013). This was followed by the development of semi-dwarf, high-yielding varieties of rice and hybrids of maize, sorghum and pearl millet. These varieties/ hybrids of cereals needed 100–150 kg N ha^{-1} in one growing season, and this could be met only by urea or other N fertilizers; organic manures had no chance and their use became almost nil over a period of next 25 years or so. Well-off young farmers cannot even think of making manure heaps or compost pits, the job is too hard and not a clean one. However, what is ignored is the fact, that organic manures not only supply plant nutrients but also build up soil organic matter, which improves soil physical properties, helps in storing moisture and is the elixir of life for the soil biota (Lal, 2014). A large numbers of reports are available on the effect of adding organic manures and crop residues on improving soil physical (Hati *et al.*, 2007; Brar *et al.*, 2015) and chemical and biological properties of soil (Prasad *et al.*, 1999; Prasad and Mishra, 2001; Sharma *et al.*, 2009; Prasad *et al.* 2014). Thanks for the organic farming (Prasad, 2005), the interest in organic manures and composts has increased in India and it is now a sought after material. However, it has not entered the main stream of crop production.

There are three major kinds of raw materials for making organic manures. The materials and tappable amounts of nutrients ($N + P_2O_5 + K_2O$) that they can provide by 2025 as estimated by Tandon (1997) are:

1. Livestock dung – estimated to provide 2.26 million metric tonnes (MMT) of (N + P_2O_5 + K_2O).
2. Crop residues – estimated to provide 3.39 MMT of (N + P_2O_5 + K_2O).
3. Human excreta – estimated to provide 2.1 MMT of (N + P_2O_5 + K_2O).

Out of the above three raw materials, only the first two, namely livestock dung and crop residues, are of importance and are discussed. Human excreta is being used only on a limited scale as sewage sludge/ water in peri-urban agriculture in growing vegetables and flowers in the neighbourhoods of some major cities and nutrient toxicity with heavy metals is a matter of concern (Saha *et al.*, 2017).

ANIMAL DUNG AND URINE

Animal dung and urine have been the materials used for replenishing soil fertility since ages, throughout the world and even today these play an important role, especially in dairy centered agriculture countries, such as, Netherlands (Yan *et al.*, 2017). In Indian agriculture their use in improving soil fertility is mentioned in *Atharva Veda* (BC 321–186) and *Agnipurana* (500–700 AD) (Raychaudhuri, 1984). Animal dung and urine with crop residues are the raw materials for making different composts. The NPK composition of dung and urine of different animals is given in Table 14.1.

Table 14.1: Nitrogen, phosphorus and potassium content of animal dung and urine

Material	N (%)	P_2O_5 (%)	K_2O (%)
Cattle dung and urine mixed	0.60	0.15	0.45
Cattle urine	0.9–1.2	Traces	0.5–1.0
Sheep dung	0.95	0.35	1.0
Sheep urine	1.6–1.7	Traces	1.8–2.0
Poultry manure	1.0–1.8	1.4–1.8	0.8–0.9
Horse dung	0.70	0.25	0.55
Horse urine	1.2–1.5	Traces	1.3–1.5

Source: ICAR (2006)

Of the above, poultry manure is the richest in NPK and is in demand. There is a practice of penning the sheep overnight in the field in some regions of the country for improving the soil fertility (*leisaindia@yahoo.co.in*).

COMPOSTING

Most crop residues have a wide C : N ratio varying from 50–120 or even more and if incorporated in soil, lead to temporary immobilization of native or applied nitrogen, and therefore these should be applied 2–3 months in advance of sowing/ planting a crop. Composting is a process by which crop residues in rural areas and kitchen or other organic trash from households is decomposed in specially made pits to reduce the C : N ratio of the final product (compost) to 10–15:1. During 2010–11, 102 million metric tons (tonnes) (MMT) of compost was produced in India, which was applied to 18.6 million ha of land and is estimated to have contributed 2 MMT of primary plant nutrients (N + P_2O_5 + K_2O) (Manna *et al.*, 2012). Broadly speaking there are two kinds of composts, namely rural compost and urban compost, the latter being produced in the largest amount (about 90% of the total). There are a number of composting processes which, could be anaerobic or aerobic, the latter being about 85–90% of the total compost produced. The major difference between the two kinds of processes is that of aeration, which is done by periodic stirring/mixing of the material being composted.

Anaerobic Composting

Anaerobic composting is generally done for urban compost, such as, in Bangalore process developed by Dr C.N. Acharya. One meter deep pits of a desirable width and length are made and are filled with about 15 cm thick layers of kitchen or other city organic trash followed by a 5 cm layer of human excreta or a mixture of sand and urine. This sequence of layering is followed until the pit is filled 30 cm above the ground level. Composting is done for a period of 5–6 months. The major problem in anaerobic composting is undesirable odour, which is produced due to production of methane, organic acids, hydrogen sulfide etc. Manure pits have therefore to be made in far off places outside the towns.

Aerobic Composting

Most composting process, especially of rural compost is aerobic and involves regular stirring for aeration. Also in rural compost the main ingredients are crop residues and cattle/ sheep/ horse/ poultry excreta. There are a number of processes, such as, Indore process, Mayadas process, Coimbatore process. These processes differ in size of pits, layering of various materials and stirring or mixing schedules. *For details,* reference may be made to Misra *et al.* (2003) and Kumar *et al.* (2014).

Accelerated Composting

Time taken for composting can be reduced by inoculating compost pit with the cultures of cellulose decomposers *Aspergillus* sp., *Penicillium* sp. and *Trichoderma viride* (Gaur *et al.*, 2002). This technology is being used by commercial compost producers.

Vermicompost

The compost made with the help of earthworms has really attracted those practicing organic farming. Earthworms consume all kind of organic matter and after extracting part of nutrients for their body build-up, the rest is excreted in a fine state in earthworm castings. The first role of earthworms is to breakdown the organic residues to fine particles, which accelerates their decomposition. The second role of earthworms is aeration by making burrows in the compost pile. The third role is enriching the compost by their own tissue. Vermicompost is thus a mixture of decomposed organic residues including humus, earthworm cocoons and live earthworms (Fig. 14.1). Vermicompost is highly priced and sought after. During 2010–11, 11.4 MMT vermicompost was made in India (Manna *et al.*, 2012). Vermicompost is richer in NPK than FYM and general compost. Some data on its composition vis-à-vis farmyard manure (FYM) are given in Table 14.2.

Table 14.2: Relative nutrient composition of vermicompost and farmyard manure

Nutrient	Vermi-compost	Farmyard manure	General compost
N (%)	2.1–2.6	1.1–1.5	1.2–1.5
P (%)	1.5–1.7	0.7–0.8	0.7–0.9
K (%)	1.4–1.6	0.6–0.7	0.6–0.7

Source: Rana, S.S., Biological Intensive Nutrient Management (BINM) (via internet)

In a study at Sriniketan (West Bengal) on wet season rice, application of vermicompost (1.6% N, 1% P and 1% K) @ 10 t ha^{-1} along with 50% recommended dose of fertilizer (RDF) was superior to 100% RDF (60 kg N + 30 kg P_2O_5 + 30 kg K_2O) and produced 16% more rice, while an application of FYM (0.5% N, 0.2% P and 0.5% K) @ 10 t ha^{-1} along with 50% RDF was at par with 100% RDF (Barik *et al.*, 2006). Thus, vermicompost was superior to FYM.

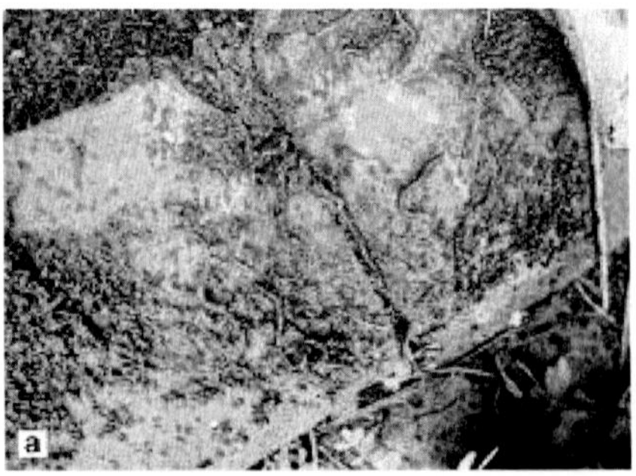

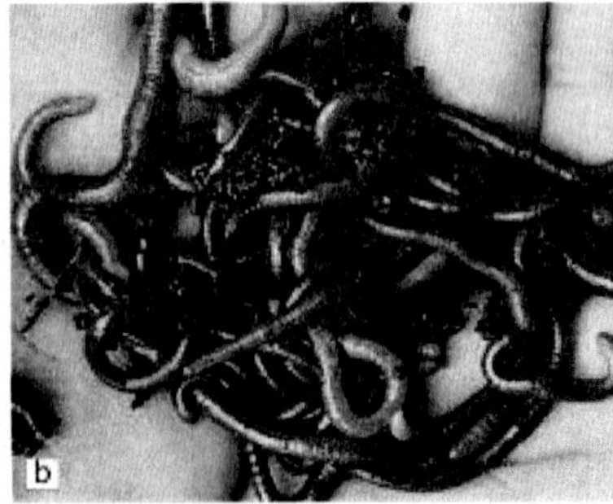

Fig.14.1: A specimen (a) of vermicompost made using *Eisenia foetida* (b)

Source: S.S. Rana, Biological Intensive Nutrient Management (BINM)-Vermicompost (via internet)

Enriched Compost

Compost can be enriched with N and P by incorporating ground phosphate rock and cultures of (N-fixing bacteria) *Azotobacter chroccoccum* Beijerinck and P-solubilizing organisms, namely *Aspergillus awamori* Nakaz. or *Bacillus polymyxa* (Prazmowski) (Gaur *et al.*, 2002).

BIOGAS SLURRY MANURE

Biogas is a useful solution to generate energy at the rural level operative at small scale and targeted for the farmers who mostly practice dairying along with farming. Biogas units produce methane, which can be used for cooking in the kitchen, lighting the house and even for running tube well motors. It also prevents the burning of cowdung in the kitchen after making cowdung cakes, a common practice in rural India. Interestingly the slurry left after the production of methane from the biogas units is much richer manure than cow dung (Awasthi, 1975). India has about 2 million biogas plants (Fig. 14.2) and it is subsidized by the Government (Tomar, 1995).

Fig. 14.2: A biogas plant

In rice–wheat cropping system, 70% N through urea and 30% through biogas slurry were as effective as 100% N through urea (Bharade *et al.*, 2006). Similarly at Ludhiana, application of 40% N through biogas slurry and 60% N through urea in wheat was as good as 100% N through urea (PAU, 1996).

IN-SITU CROP RESIDUE INCORPORATION (CONSERVATION AGRICULTURE)

Crop residue, especially of rice, has received special attention in the rice–wheat cropping system of northern India and Delhi has been facing a serious health problem, especially for asthmatic elderly persons and children, due to burning of straw in Punjab and Haryana (Prasad and Shivay, 2018). Burning of rice straw emits carbon dioxide (CO_2), sulphur dioxide (SO_2), nitric oxide (NO_2) and particulate matter (PM) of different sizes varying from 2.5 to 10 nm (Agarwal *et al.*, 2012), which are harmful to lungs and trigger bronchial asthma (Torigoe *et al.*, 2000). In addition, burning of straw also leads to loss of plant nutrients N, P and S. Singh *et al.* (2008) reported that, by burning of rice straw 35 kg N, 3 kg P (7 kg P_2O_5) and 2.7 kg S is lost per hectare each year in Punjab.

In conservation agriculture as practiced in the USA, crop residues are incorporated in soil to improve soil fertility and to store moisture (Prasad and Power, 1991). In India also, a large number of studies have been made at the Punjab Agricultural University (PAU), Ludhiana (Yadvinder-Singh *et al.*,2005; Singh *et al.*, 2008; Jat *et al.*, 2014), Indian Agricultural Research Institute (IARI), New Delhi (Prasad *et al.*, 1999) and elsewhere (Gangwar *et al.*, 2006) on the *in-situ* incorporation of rice straw after the harvesting of rice crop. Incorporation of rice straw showed some N deficiency in the early stages of wheat growth and extra N application of 25–30 kg ha^{-1} had to be made to overcome this (Gangwar *et al.*, 2006; Bijay–Singh *et al.*, 2008). Most studies showed that, yields with incorporation of rice straw were comparable with those obtained after straw burning or straw removal and soil fertility was improved. It may be added that while rice straw contains about 40–50% of the total N and P removed by the crop, it contains 93% of Ca, 87–89% of K and Mg, 69% of S, 85% of Zn, 83% of Fe, 93% of Mn and 95% of Cu removed by the crop (Lakshmanan and Prasad, 2004; Lakshmanan *et al.*, 2005; Shivay *et al.*, 2014) and thus *in-situ* incorporation of rice straw can reduce the need for application of these nutrients, specially, S and Zn in these soils.

SUMMARY

1. Organic manures improve soil physical, chemical and biological properties.
2. The major sources of organic manures are: cattle, sheep and poultry excreta/droppings, crop residues and human excreta.
3. Human-excreta are utilized on a limited scale as sewage-sludge and water in peri-urban agriculture.
4. In conservation agriculture, crop residues are retained on soil surface. This has met with limited success in recycling rice straw in rice–wheat cropping system.
5. The best way to use the animal excreta and crop residue is by making compost.
6. The major objective in making compost is to reduce the C : N ratio in crop residues from 50–120 or even more to 10–15.
7. Composting processes could be aerobic or anaerobic. The major difference is that in aerobic processes, the manure is stirred or mixed

periodically to provide aeration. Of these, aerobic processes are mostly used.

8. Vermicompost is made by employing earthworms.
9. Compost can be enriched with ground phosphate rock and incubation with cultures of N- fixing *Aztobacter* sp. and P-solubilizing P-organisms *Aspergillus awamori* or *Paenibacillus polymyxa*.
10. Cowdung can be used for producing biogas (methane), which can be used for cooking, lighting or running tube-well motors.
11. Biogas slurry is good quality manure.

REFERENCES

Agarwal, R., Awasthi, A., Singh, N., Gupta, P.K. and Mittal, S.K. 2012. Effect of exposure to rice crop residue burning smoke on pulmonary functions and oxygen saturation level of human beings in Patiala (India). *Science of the Total Environment* **429**: 161–166.

Awasthi, R.K. 1975. Bio gas (gobar gas) plant in perspective. *Financing Agriculture* **7**(3): 1–7.

Barik, A.K., Das, A., Giri, A.K. and Chattopadhyay, G.N. 2006. Effect of integrated plant nutrient management on growth, yield and production economics of wet season rice (*Oryza sativa*). *Indian Journal of Agricultural Sciences* **76**(11): 657–663.

Bharade, N.M., Shivay, Y.S., Singh, S. and Kumar, D. 2006. Productivity and economics of rice (*Oryza sativa*)–wheat (*Triticum aestivum*) cropping system as influenced by neem oil-treated urea and biogas slurry. *Indian Journal of Agricultural Sciences* **76**(9): 550–553.

Bijay-Singh, Shan, Y.H., Johnson-Beebout, S.E. and Yadvinder-Singh. 2008. Crop residue management for lowland rice-based cropping systems in Asia. *Advances in Agronomy* **98**: 118–199.

Brar, B.S., Singh, J., Singh, G. and Kaur, G. 2015. Effects of long term application of inorganic and organic fertilizers on soil organic carbon and physical properties in maize–wheat rotation. *Agronomy* **5**: 220–238.

Gangwar, K.S., Singh, K.K., Sharma, S.K. and Tomar, R.K. 2006. Alternative tillage and crop residue management in wheat after rice in sandy loam soils of Indo-Gangetic plains. *Soil and Tillage Research* **88**: 242–252.

Garbyal, S.S. 1999. *Jhuming* (shifting cultivation) in Mizoram (India) and new land policy – how far has succeeded in containing the primitive agricultural practice. *Indian Forester* **125**(2): 137–148.

Gaur, A.C., Neelkantan, S. and Dargan, S. 2002. *Organic Manures*, 159 pp. Indian Council of Agricultural Research, New Delhi.

Hati, K.M., Swarup, A., Dwivedi, A.K., Misra, A.K. and Bandyopadhyay, K.K. 2007. Changes in soil physical properties and organic carbon status at the topsoil horizon of a Vertisol of central India after 28 years of continuous cropping, fertilization and manuring. *Agriculture, Ecosystems and Environment* **119**(1–2): 127–134.

ICAR. 2006. Handbook of Agriculture, pp. 429. Indian Council of Agricultural Research, New Delhi.

Jat, M.L., Yadvinder-Singh, Sidhu, H.S., Parvinder-Singh and Jat, R.K. 2014. Managing cereal crop residues for sustainable crop production systems. (In) *Textbook of Plant Nutrient Management,* pp. 326–347. Prasad, R., Kumar, D., Rana, D.S., Shivay, Y.S. and Tewatia, R.K. (Eds.), Indian Society of Agronomy, New Delhi.

Kumar, D., Singh, G. and Prasad, R. 2014. Organic manures. (In) *Textbook of Plant Nutrient Management,* pp. 268–294. Prasad, R., Kumar, D., Rana, D.S., Shivay, Y.S. and Tewatia, R.K. (Eds.). Indian Society of Agronomy, New Delhi.

Laksmanan, R. and Prasad, R. 2004. Concentration and uptake of secondary nutrients (Ca, Mg, S) in rice as influenced by duration of variety and nitrogen fertilization. *Acta Agronomica Hungarica* **52**: 133–139.

Laksmanan, R., Prasad, R. and Jain, M.C. 2005. Yield and uptake of micronutrients by rice as influenced by variety and nitrogen fertilization. *Archives of Agronomy and Soil Science* **51**: 1–14.

Lal, R. 2014. Soil carbon sequestration for sustained nutrient supply. (In) *Textbook of Plant Nutrient Management,* pp: 48–63. Prasad R., Kumar D., Rana D.S., Shivay Y.S. and Tewatia R.K. (Eds). Indian Society of Agronomy, New Delhi.

Manna, M.C., Subba Rao, A., Sahu, A. and Singh, U.B. 2012. Compost availability and its impact on crop productivity and soil health. (In) *Compost Handbook: Research – Production – Application.* Fertilizer Development and Consultation Organization (FDCO), New Delhi. 132 pp.

Misra, R.V., Roy, R.N. and Hiraoka, H. 2003. *On-farm Composting Methods*, Food and Agriculture Organization (FAO), Rome, 34 pp.

PAU. 1996. AICRIP on Renewable Energy Sources of Agriculture and Agro-based Industry, Department of Engineering, Punjab Agricultural University, Ludhiana.

Prasad, R. 2005. Modern agriculture vis-à-vis organic farming. *Current Science* **89**: 252–254.

Prasad, R. and Mishra, B.N. 2001. Effect of addition of organic residues, farmyard manure and fertilizer nitrogen on soil fertility in rice–wheat cropping system. *Archives of Agronomy and Soil Science* **46**: 455–463.

Prasad, R. and Power, J.F. 1991. Crop residue management. *Advances in Soil Science* **15**: 205–251.

Prasad, R. and Shivay, Y.S. 2018. Management options to alleviate the menace of rice straw burning – An overview. *Indian Journal of Agricultural Sciences* **88**(11): 1,651–1,660.

Prasad, R., Gangaiah, B. and Aipe, K.C. 1999. Effect of crop residue management in a rice–wheat cropping system on growth and yield of crops and soil fertility. *Experimental Agriculture* **35**: 427–435.

Prasad, R., Singh, G. and Prasad, R. 2014. Organic manures. (In) *Textbook of Plant Nutrient Management,* pp: 268–294. Prasad, R., Kumar, D., Rana, D.S., Shivay, Y.S. and Tewatia, R.K. (Eds), Indian Society of Agronomy, New Delhi.

Raychaudhuri, S.P. 1984. Development of soil productivity in ancient and medieval India. (In) *Soil Science in India, Indian Society of the Soil Science*, New Delhi, Bulletin No. **14**: 1–13.

Saha, S., Saha, B.N., Pati, S., Pal, B. and Hazra, G.C. 2017. Agricultural use of sewage sludge in India: benefits and potential risk of heavy metals contamination and possible remediation options – a review. *International Journal of Environmental Technology and Management* **20**(3–4): (https://doi.org/10.1504/ IJETM.2017.089645).

Sharma, S.N., Prasad, R., Dwivedi, M.K., Kumar, S., Davari, M.R. and Ram, M. 2009. Effect of cropping system on production and chemical and biological properties of soil. *Archives of Agronomy and Soil Science* **55**: 429–438.

Shivay, Y.S., Prasad, R. and Pal, M. 2014. Effect of levels and sources of sulfur on yield, sulfur and nitrogen concentration and uptake and S-use efficiency in *Basmati* rice. *Communications in Soil Science and Plant Analysis* **45**(18): 2,468–2,479.

Singh, R.P., Dhaliwal, H.S., Sidhu, H.S., Manpreet-Singh, Y.S. and Blackwell, J. 2008. Economic assessment of the Happy Seeder for rice–wheat systems in Punjab, India. *Conference Paper, AARES 52nd Annual conference, Canberra.* Australia: ACT.

Swaminathan, M.S. 2013. Genesis and growth of the yield revolution in wheat in India: Lessons from shaping our agricultural destiny. *Agricultural Research* **2**(3): 183–188.

Tandon, H.L.S. 1997. Organic resources: An assessment of potential supplies, their contribution to agricultural productivity, and policy issues of Indian agriculture from 2000–2025. (In) *Plant Nutrient Needs, Supply, Efficiency and Policy issues: 2000–2025*. Kanwar, J.S. and Katyal, J.C. (Eds.), National Academy of Agricultural Sciences, New Delhi, India.

Tandon, H.L.S. 2014. History of soil fertility. (In) *Textbook of Plant Nutrient Management,* pp. 1–22. Prasad, R., Kumar, D., Rana, D.S., Shivay, Y.S. and Tewatia, R.K. (Eds.). Indian Society of Agronomy, New Delhi.

Tomar, S.S. 1995. Status of biogas plants in India–An overview. *Energy for Sustainable Development* **1**(5): 53–56.

Torigoe, K., Hasegawa, S., Numata, O., Yazaki, S., Matsunaga, M., Boku, N., Hiura, M. and Ino, H. 2000. Influence of emission from rice straw burning on bronchial asthma in children. *Pediatrics International* **42**(2): 143–150.

Yadvinder-Singh, Bijay-Singh and Timsina, J. 2005. Crop residue management for nutrient cycling and improving soil productivity in rice-based cropping systems in the tropics. *Advances in Agronomy* **85**: 269–407.

Yan, J., de Buisonje, F.E. and Melse, R.W. 2017. White paper on livestock manure treatment technology of Netherlands and situation of China. *Wageningen Livestock Research Report* 1048. (https:/ doi.org/10.18174/ 423982).

QUESTIONS

1. Write short notes on

(a) *In-situ* incorporation of crop residues.
(b) Bio-gas
(c) Vermicompost
(d) Peri-urban agriculture.
(e) Enrichment of compost.

2. Match the followings

(a) Night soil	(i) *Eisenia foetida*
(b) Particulate matter of 2.5–10 nm	(ii) N-fixing bacteria
(c) *Trichoderma viride*	(iii) P-solubilization
(d) *Azotobacter* sp.	(iv) Anaerobic composting
(e) Vermicompost	(v) Asthma

Chapter 15

Legumes and Soil Fertility

Soil-rejuvenation power of legumes in crop rotations has been known in several parts of the world since ages. It is mentioned in the Indian agriculture of the Chalcolithic period (*ca* 1000 BC) (Raychaudhuri and Mira, 1993), in Chinese agriculture in the Han dynasty (*ca* 2nd century) (Hsu, 1980) and in Roman agriculture (White, 1970). Legumes can be fitted in cropping systems in India in a number of ways, such as, a crop in a cropping system, a catch crop in an intensive cropping system, an intercrop and as a green manure. A brief discussion on these follows.

LEGUME AS A MAIN CROP IN A CROPPING SYSTEM

There are a large number of traditional cropping systems that include a legume. Some of these are listed in Table 15.1.

In recent years, growing of chickpea in rice–fallows in eastern and southern states has received considerable attention (Mula *et al.*, 2015; Nagoli *et al.*, 2015; Chandrakar *et al.*, 2017) owing to higher yield, market demand and profit. When included in a cropping system, grain legumes can fix 50–150 kg N ha^{-1} (see Table 6.1) (Vyas *et al.*, 2014) and leave about 25–50 kg N ha^{-1} to the succeeding non-legume crop; to translate it in a common man's language this amounts to a saving of at least one *katta* (45 kg bag) of urea for the succeeding crop. Forage legumes, such as, alfalfa can fix about 300 kg N ha^{-1}.

In addition to N benefits, legumes also improve the availability of other nutrients. Legumes owing to their root-system improve soil structure, porosity, biological activity and recycle plant nutrients (USDA-NRCS-SQI, 1998). Copeland and Crookston (1992) reported that, K and total micronutrient content increased in corn grown in a 2-year rotation with soybean as compared with continuous corn. These improvements in soil health result in an increase in the yield of the crop succeeding a legume. For example, Ahlawat *et al.* (1998) reported that wheat yields after *kharif* soybean, pigeonpea or mungbean were 1.3–1.6 t ha^{-1} higher than wheat following *kharif* pearl millet and sorghum.

LEGUME AS A CATCH CROP

Legume can be grown as a catch crop between two crops in a cropping system. Growing of summer mungbean after wheat harvesting and incorporation of its residue after taking a picking of pods before rice transplanting in the rice–wheat cropping system (RWCS) has received considerable attention in the recent years (Sharma *et al.*, 2000, 2009; Sekhon *et al.*, 2004, 2007; Prasad, 2016). Rice yield owing to incorporation of summer mungbean residue after picking the first flush of pods was 0.6–0.8 t ha^{-1} higher than that after fallow, and fertilizer N equivalent for the residual effect of this practice was 60–90 kg N ha^{-1} in the RWCS (Sharma and Prasad, 1999). In addition, this practice improved the soil physical, chemical and biological properties. Mandal *et al.* (1999) reported that incorporation of summer mungbean residue increased mean weight diameter of aggregates by 38–74% after rice harvesting and 30–37% after wheat harvesting and percentage of water-soluble aggregates by 12–14% after rice harvesting and 14% after wheat harvesting in the surface 0–15 cm soil. Sharma *et al.* (2009) reported that, microbial biomass and microbial activity as judged by CO_2 evolution and cell count for bacteria, fungi and actinomycetes was significantly higher due to mungbean residue incorporation.

Table 15.1. Traditional legume–based cropping systems in different agro-climatic regions of India

Sl. No.	Agro-climatic zone	States represented	Annual rainfall (mm)	Cropping system (*Kharif–rabi*–summer)
1	Western Himalayan Region	Jammu and Kashmir Himachal Pradesh, Uttarakhand	1,650–2,000	Rice–chickpea/ lentil/ fieldpea, Maize–chickpea/ fieldpea/ lentil, Pigeonpea–wheat, Mungbean/ urdbean–mustard, Common bean–potato
2	Eastern Himalayan Region	Asom, Bengal, Manipur, Meghalaya, Nagaland, Arunachal Pradesh	1,840–3,530	Early rice–urdbean/ mungbean, Rice–lathyrus, Maize–maize–urdbean, Maize–pigeonpea/ horsegram, Maize–chickpea/ lentil/ fieldpea, Jute–urdbean–chickpea/ lentil
3	Lower Gangetic Plains Region	West Bengal	1,300–1,600	Maize–chickpea/ lentil/ fieldpea, Rice–chickpea/ lentil/fieldpea, Rice–chickpea + mustard/lentil

4	Middle Gangetic Plains Region	Uttar Pradesh, Bihar	1,200–1,470	Maize–wheat–summer mungbean/urdbean, Rice–potato–summer mungbean/urdbean, Rice–chickpea/lentil
5	Upper Gangetic Plains Region	Uttar Pradesh	720–980	Rice–wheat/potato–summer mungbean, Maize–wheat/potato–summer mungbean, Mungbean/ urdbean–wheat, Sorghum (fodder)–chickpea
6	Trans Gangetic Plains Region	Punjab, Haryana	360–890	Maize–potato–mungbean/urdbean, Rice/maize–wheat–mungbean/urdbean, Maize–potato–mungbean/urdbean, Rice–chickpea/lentil, Maize–chickpea/lentil/fieldpea
7	Eastern Plateau and Hills Region	Madhya Pradesh, Maharashtra, Odisha, West Bengal	1,270–1,430	Early rice–urdbean, Rice–rice–cowpea, Jute–cowpea/urdbean, Maize–cowpea
8	Central Plateau and Hill Region	Madhya Pradesh, Rajasthan, Uttar Pradesh	490–1,570	Sorghum (grain/fodder)–chickpea, Fallow–chickpea, Sorghum + pigeonpea, Pearlmillet + pigeonpea, Rice–chickpea/lentil/fieldpea, Maize–chickpea/lentil/fieldpea, Mothbean/mungbean/urdbean–wheat, Pearlmillet–chickpea
9	Western Plateau and Hill Region	Maharashtra, Madhya Pradesh, Rajasthan	600–1,040	Urdbean–*rabi* sorghum, Sorghum–potato–mungbean, Cotton + urdbean/mungbean–fallow, Sorghum–wheat–cowpea/mungbean, Cotton–chickpea, Sorghum–chickpea, Mungbean/urdbean–safflower
10	Southern Plateau and Hill Region	Andhra Pradesh, Tamil Nadu, Karnataka	680–1,000	Maize–sorghum + pigeonpea, Sorghum–chickpea, Pearlmillet–horsegram, Mungbean/urdbean–safflower, Rice–mungbean/urdbean/cowpea, Mungbean–sorghum/safflower, Mungbean–pigeonpea, Rice–rice +

				mungbean/urdbean/cowpea
11	East coast Plains and Hills Region	Odisha, Andhra Pradesh, Tamil Nadu, Puducherry	780–1,290	Rice–mungbean/urdbean, Sorghum–mungbean/ urdbean, Tapoica + mungbean/ urdbean, Rice–rice, Rice–maize/cowpea, Maize–horsegram/ pigeonpea/ chickpea
12	West Coast Plains and Hills Region	Tamil Nadu, Kerala, Goa, Karnataka, Maharashtra	2,230–3,640	Rice–urdbean/cowpea/ chickpea, Sugarcane + urdbean
13	Gujarat Plains and Hills Region	Gujarat	340–1,790	Urdbean–safflower/niger, Cowpea–safflower, Mungbean–tobacco, Pearlmillet + pigeonpea–chickpea, Sorghum + pigeonpea–chickpea
14	Western Dry Region	Rajasthan	400	Rice–chickpea, Cotton–chickpea, Pearlmillet–chickpea + Indian mustard, Sorghum–chickpea + Indian mustard, Mothbean/ mungbean–wheat
15	Island Region	Andaman and Nicobar islands, Lakshwadeep islands	1,500–3,090	Rice–rice–urdbean, Maize–rice–urdbean

Source: Singh *et al.* (2005)

LEGUME AS AN INTERCROP

Legumes are grown as an intercrop in widely spaced crops, such as, maize, sorghum, pearl millet, cotton, sugarcane, pigeonpea, castor (*Ricinus communis* L.) etc. all over the country (Table 15.2). Short-duration legumes, such as, mungbean and urdbean are the most common choice. However, farmers practicing dairying and those near towns prefer cowpea [*Vigna unguiculata* (L.) Walp.], from which they harvest green pods for selling as a green vegetable and using the green foliage as protein rich green fodder for milch cows. Since legumes generally have indeterminate fruiting, harvesting of green pods can be done over a stretch of time. For grains, mungbean and urdbean are harvested at maturity. There are six advantages of intercropping with legumes. These are: *1*, additional produce from the cropping system; *2*, additional income from the cropping system; *3*, mid-season income before the harvesting of main crop; *4*, suppression of weeds; *5*, some contribution of N to the main crop; *6*, increased efficiency of other nutrients applied to the cropping system. Ali (1988) reported that, in pigeonpea–legume intercropping system, the weed suppression ability for cowpea, urdbean (blackgram) and mungbean was 39.1, 32.8 and 31.0%, respectively; in general it was related to their canopy and foliage. Senaratne *et al.* (1995) using ^{15}N technique reported that in maize–legume intercropping systems, maize derived 7–11% of its N from mungbean, 11–20% from cowpea and 12–26% from groundnut. Yadav (1986) reported that, efficiency of P applied to sugarcane increased from 3.5% in sole cropping to 15.8% when sugarcane was intercropped with mungbean.

Table 15.2: Some legume–based intercropping systems in different states of India

Cropping system	Row ratio	States
Maize + mungbean/ urdbean/ cowpea (fodder)	1 : 2	All maize-growing states
Sorghum + mungbean/ urdbean/ cowpea (fodder)	1 : 2	Maharashtra, Madhya Pradesh (MP)
Pearl millet + mungbean/ urdbean/ cowpea (fodder)	1 : 2	Rajasthan
Pigeonpea + mungbean/ urdbean	2 : 1	Maharashtra, Madhya Pradesh, Karnataka
Pigeonpea (long duration) + groundnut	1 : 2	Central Uttar Pradesh
Castor + mungbean/ urdbean	1 : 3	Gujarat, Odisha, Tamil Nadu, Telangana
Spring sugarcane + mungbean/ urdbean	1 : 2	West Bengal, Uttar Pradesh, Bihar, Haryana
Jute (seed crop) + urdbean	1 : 1	West Bengal, Odisha, Asom
Cotton + mungbean/ urdbean	1 : 2	Andhra Pradesh, Maharashtra, Tamil Nadu
Spring sunflower + mungbean/ urdbean	1 : 3	Uttar Pradesh, Haryana, Punjab, Bihar

Modified from Vyas *et al.* (2014)

LEGUME AS A GREEN-MANURE

Green-manuring with legumes is an age-old practice and is practiced more in rice-growing regions of India, especially in the south. The common practice is to grow a legume before the rice season and incorporate it in soil before preparing the land for transplanting rice. Some farmers grow green-manure plants on the bunds and incorporate them in field before preparing the land for rice transplanting. Alternatively, a legume can be sown along with rice and allowed to grow for about 30 days and then knocked down by 2,4-D @ 500 g ha^{-1}; this practice is known as brown manuring. Still another way of green-manuring is lopping the green forest tree or shrub leaves and twigs and incorporating these in soil before rice transplanting. Some of the species, whose leaves and twigs are used as green manure, are: *awaram* [*Senna auriculata* (L.) Roxb.; syn. *Cassia auriculata* L.], sickle senna [*Senna tora*

(L.) Roxb.], *Cassia tora* (L.), coffee senna [*Senna occidentalis* (L.) Link.]; syn. *Cassia occidentalis* (L.) Rose, *Gliricidia maculata,* (Humb, Bonpl. & Kunth) Sleud., river tamarind or *Subabul* [*Leucaena leucocephala* (Lam.) lewit], crown flower [*Calatropis gigantea* (L.) W.T. Aiton], wild indigo [*Tephrosia punpurea* (L.) Pers.] and white hoarypea [*Tephrosia candida* (Roxb.) DC].

A list of legumes used for green manuring along with their NPK composition is given in Table 15.3.

Table 15.3: Common green-manure legumes for green-manuring

Green manure	Biological name	% on fresh weight basis			N accumulated at 60 days after sowing (kg ha^{-1})
		N	P_2O_5	K_2O	
Dhaincha	*Sesbania bispinosa* (Jacq.) W. Wight	0.62	–	–	133
Sunnhemp	*Crotolaria juncea* L	0.75	0.12	0.50	134
Phillipesara	*Phaseolus trilobus* (L) Aiton	–	–	–	102
Cowpea	*Vigna unguiculata* L	0.71	0.15	0.58	74
Mungbean (Greengram)	*Vigna radiata* (L) R. Wilczek	0.72	0.18	0.53	–
Urdbean (Blackgram)	*Vigna mungo* (L) Hepper	0.85	0.18	0.53	–
Cluster bean	*Cyamopsis tetragonoloba* (L) Taub	0.34	–	–	91

Source: Modified from Buttar and Rana (2014)

SUMMARY

1. Legumes have been used since ages for rejuvenating soil.
2. When included in a cropping system, they provide residual N and due to their deep roots recycle other nutrients including micronutrients.
3. Legumes improve physical and biological properties of soils.
4. Legumes are ideal for rice fallows. In this connection, chickpea has received special attention.
5. When grown as an intercrop, they increase the productivity and profit and supply some N to the main crop.
6. As a catch crop in an intensive cropping system, they increase productivity, profit and improve soil physical, chemical and biological properties.
7. Common legume green manuring crops are: mungbean, urdbean, cowpea, sunnhemp, *Phillipesara* and clusterbean.
8. The trees whose leaves and twigs are used as green manures include: *Cassia auriculata, Senna tora, Senna occidentalis, Gliricidia maculata, Leucaena leucocephala, Calatropis gigantea, Tephrosia purpurea* and *Tephrosia candida.*

REFERENCES

Ahlawat, I.P.S., Ali, M., Yadav, R.L., Kumar Rao, J.V.D.K., Johansen, C. and Rego, T.J. 1998. Biological nitrogen fixation and residual effects of summer and rainy season grain legumes in rice and wheat cropping systems of the Indo-Gangetic Plain. (In) *Residual Effects of Legumes in Rice and Wheat Cropping Systems of the Indo-Gangetic Plain.* Kumar Rao, J.V.D.K., Johansen, C. and Rego, T.J. (Eds). International Crops Research Institute for the Semi-Arid tropics, Patancheru, Hyderabad, Telangana, pp. 31–54.

Ali, M. 1988. Weed suppressing ability and productivity of short-duration legumes intercropped with pigeonpea under rainfed conditions. *Tropical Pest Management* **34**(4): 384–387.

Buttar, G.S. and Rana, D.S. 2014. Legume as green manure. (In) *Textbook of Plant Nutrient Management,* pp. 311–325. Prasad, R., Kumar, D., Rana, D.S., Shivay, Y.S. and Tewatia, R.K. (Eds.), Indian Society of Agronomy, New Delhi.

Chandrakar, C.K., Bhambri, M.C., Pali, G.P., Kumar, S., Jangde, A., Pandey, K.K. and Singh, S. 2017. Response of plant nutrients on soil fertility, productivity and profitability of rice (*Oryza sativa*)–chickpea (*Cicer arietinum*) in Chhattisgarh plains. *International Journal of Current Microbiology and Applied Sciences* **6**(4): 1,867–1,875.

Copeland, P.J. and Crookston, P.K. 1992. Crop sequence affects nutrient composition of corn and soybean grown under high fertility. *Agronomy Journal* **84**: 503–509.

Hsu, C. 1980. *Han Agriculture*, 377 pp. University of Washington Press, Seattle, WA, USA.

Mandal, U.K., Sharma, S.N., Singh, G. and Das, D.K. 1999. Effect of *Sesbania* green manuring and greengram (*Vigna radiata*) residue incorporation on physical properties of soil in rice (*Oryza sativa*)–wheat (*Triticum aestivum*) cropping system. *Indian Journal of Agricultural Sciences* **69**(9): 615–620.

Mula, M.G., Samineni, S. and Gaur, P.M. 2015. *Chickpea in the Rice–Fallow Cropping System.* ICRISAT– Government of Odisha, Patancheru, Hyderabad, Telangana.

Nagoli, S.B., Basavaneppa, M.A., Sawarangaonkar, G.L., Biradar, S.A., Tevari, P. and Navyashree, M.R. 2015.

Soil properties as influenced by alternate cropping systems to rice in Tung Bhadra Project area. *International Journal of Technology and Agriculture* **33**(4): 2,481–2,484.

Prasad, R., Shivay, Y.S. and Chaudhari, S. 2016. Summer mungbean (*Vigna radiata*) for alleviating deficit of pulses in India. *Indian Journal of Agronomy* **61**(4th IAC Special issue): S106–S112.

Raychaudhuri, S.P. and Mira, R. 1993. *Agriculture in Ancient India–A report.* Indian Council of Agricultural Research, New Delhi, India, 201 pp.

Sekhon, H.S., Bains, T.S., Kooner, B.S. and Sharma, P. 2007. Grow summer mung for improving crop sustainability, farm income and malnutrition. *Acta Horticulturae* **752**: 459–464.

Sekhon, H.S., Singh, G., Star, G.S., Shanmugsundaram, S., Bains, T.S. and Konoor, B.S. 2004. *Technology package for mungbean cultivation in Punjab.* DFID-AVRDC-PAU Mungbean Project, Punjab Agricultural University, Ludhiana.

Senaratne, R., Liyanage, N. and Soper, R.J. 1995. Nitrogen fixation and transfer from cowpea, mungbean and groundnut when intercropped with maize. *Nutrient Cycling in Agroecosystems* **40**(1): 41–48.

Sharma, S.N. and Prasad, R. 1999. Effects of *Sesbania* green manuring and mungbean residue incorporation on productivity and nitrogen uptake of a rice–wheat cropping system. *Bioresource Technology* **67**: 171–175.

Sharma, S.N., Prasad, R., Dwivedi, M.K., Kumar, S., Davari, M.R. and Ram, M. 2009. Effect of cropping system on productivity and chemical and biological properties of soils. *Archives of Agronomy and Soil Science* **55**(4): 429–438.

Sharma, S.N., Prasad, R., Singh, S. and Singh, P. 2000. On-farm trials of the effect of introducing a summer green manure of mungbean on the productivity of rice–wheat cropping system. *Journal of Agricultural Sciences,* Cambridge **101**: 467–472.

Singh, G., Sekhon, H.S. and Kolar, J.S. 2005. *Pulses*. 592 pp. Agrotech Publishing Academy, Udaipur, Rajasthan. USDA-NRCS-SQI. 1998. *Legumes and Soil Quality*. USDA-NRCS-Soil Quality Institute, Auburn, AL. Soil Quality–Agronomy Note No. 6.

Vyas, A.K., Kumar, D., Ali, M., Prasad, R. and Rana, D.S. 2014. (In) *Textbook of Plant Nutrient Management,* pp: 295–310. Prasad, R., Kumar, D., Rana, D.S., Shivay, Y.S. and Tewatia, R.K.(Eds.), Indian Society of Agronomy, New Delhi.

White, K.D. 1970. Fallowing, crop rotation and crop yields in Roman times. *Agricultural History* **44**: 281–290.

Yadav, R.L. 1986. Response of sugarcane to phosphorus through legume intercropping. *Indian Journal of Sugarcane Technology* **3**: 24–28.

QUESTIONS

1. Briefly discuss

(a) Brown manuring

(b) Effect of legumes in cropping systems on physical, chemical and biological properties of soil.

(c) Tree loppings as green manure.

2. Answer True (T) or False (F)

(a) *Senna auriculata* is an excellent green-manure crop.

(b) Chickpea is recommended as a green-manuring crop for rice.

(c) Maize is an ideal crop for intercropping.

(d) Intercropping in sugarcane is done after the plants are knee-high tall.

(e) Intercropping is an age-old practice and does not qualify for modern agriculture.

Chapter 16

Making Fertilizer Recommendations Based on Soil Tests and Plant Tissue Testing

Fertilizer recommendations are generally made on the basis of soil test. Recommendations of fertilizer for top-dressing are best made on tissue testing.

SOIL TEST

A number of approaches have been adopted to increase the precision of soil-test values in making fertilizer recommendations. Four most important ones are discussed.

Low-Medium-High Approach

Making fertilizer recommendations based on soil test for a nutrient involves conducting a large number of field experiments on soils similar to those which are intended for making recommendations. For an individual nutrient (*say* P for example) experiments are conducted with graded doses of P on a crop for which the fertilizer recommendations are to be made. Thus, agronomic experiments are to be conducted for each crop on each kind of soil. The response curve can then be divided into three sections: *a*, low soil-test value and high response to fertilizer zone; *b*, medium soil-test value and medium response to fertilizer zone; and *c*, high soil-test values and no response to fertilizer zone. In India, this was done by Muhr *et al.* (1965).

Graphic Method of Determining Optimum level of Fertilization

As discussed above, in general, a graph is made by having yield on the y-axis and nutrient doses on the x-axis (Fig 16.1). However, relative yield is a better criterion than actual yield. Relative yield is the yield with optimum amounts of all nutrients except the nutrient of interest divided by the maximum yield with optimum amounts of all nutrients. Using this concept, Cate and Nelson (1965) suggested a graphic method of determining optimum dose of a nutrient (Fig 16.2). The data points in the extreme lower left and extreme upper right quadrants represent deviations from expected behaviour. Points lying in the extreme lower left quadrant represent situations where the soil-test level is below the critical level and thus a response would be expected but no response is observed. Points lying in the extreme upper right quadrant represent situations where the soil-test level is above the critical level and thus no response would be expected but a response is observed (Douglas, 2011). The data in Fig 16.2 show that soils having Olsen's available P values near 10 kg P ha^{-1} give the best yield response.

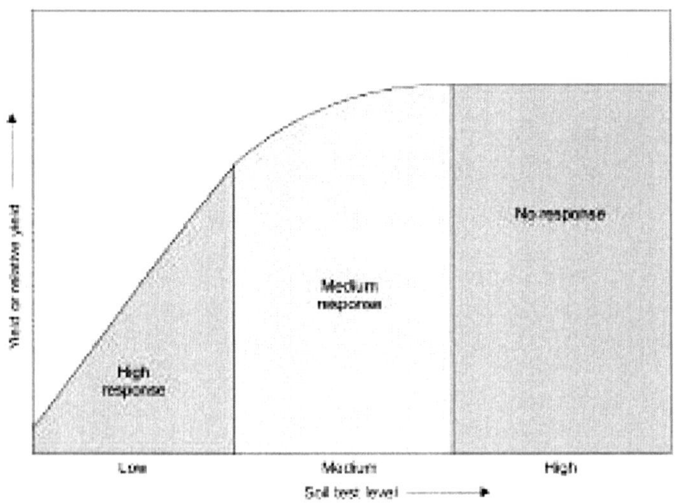

Fig.16.1: Response curve showing relationship between low, medium and high soil-test values and the kind of response expected

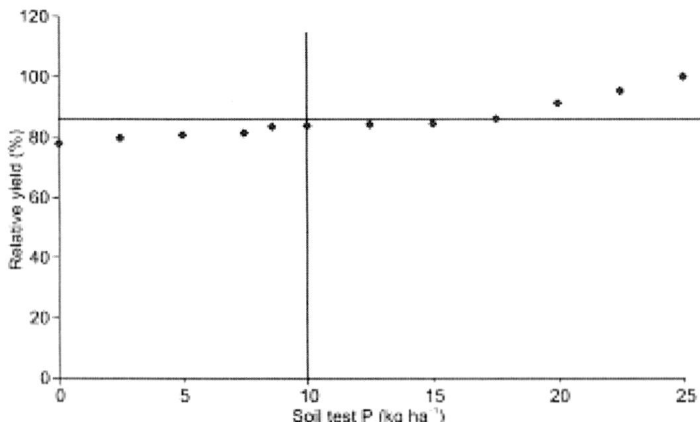

Fig. 16.2: Relative yield vs. soil-test phosphorus showing response curve and Cate-Nelson graphical separation of the data into responsive and non-responsive populations *Data source*: Jadhav (2014); Paramesh (2014); and Haokip (2014)

Targeted Yield Approach

Ramamoorthy *et al.* (1967) initiated this approach, which involves creating the following basic data for each crop, each nutrient and each location from agronomic experiments with graded dose of each nutrient. The studies were made under the ICAR-All India Coordinated Research Project on Soil-Test-Crop Response (STCR) initiated in 1967–68.

1. Nutrient requirement kg t^{-1} of grain.
2. Nutrient the soil can supply expressed as % of soil-test value for the nutrient.
3. Fertilizer-use efficiency of the nutrient.

In targeted yield (t ha^{-1}) approach for formulating fertilizer recommendations, the basic data, viz. nutrient requirement (NR) (kg t^{-1}), per cent contribution from the soil-available nutrients [CS (%)] and the per cent contribution from the applied fertilizer nutrients [CF (%)], are transferred into a workable equation (Rao and Srivastava, 2000).

Fertilizer dose = [Nutrient requirement (kg t^{-1}) of grain]/[CF (%)] × 100 × T (t ha^{-1}) – [CS (%)]/ [CF (%)] × [Soil-test value (kg ha^{-1})],

Basic data and fertilizer adjustment equations of paddy (IR 36) in alluvial soils of Barddhaman, West Bengal, are given in Table 16.1, while the data from a verification trial on a farmer's field are shown in Table 16.2 (Bera *et al.*, 2006). Data in Table 16.2 reveal that dose recommended on the basis of prediction equation almost lead to targeted yield.

Table 16.1: Basic data for equation for targeted yield for rice in alluvial soils of Barddhaman, West Bengal

Nutrient	Basic data			Targeted yield equations
	NR (kg t^{-1})	*CS* (%)	*CF* (%)	
N	14.8	13.56	67.64	$F_N = 21.9T–0.2S_N$
P_2O_5	10.5	31.59	84.67	$F_P = 12.4T–0.37S_P$
K_2O	18.6	19.33	139.89	$F_K = 13.3T–0.14S_K$

Modified from Bera *et al.* (2006)

SITE-SPECIFIC NUTRIENT MANAGEMENT (SSNM)

Soil-nutrient supplies, fertilizer efficiency, and productivity vary widely across small distances in the diverse irrigated rice fields in Asia. At present, however, blanket fertilizer recommendations are often applied over large areas without taking into account the wide variability and site- and season-specific crop-nutrient requirements within each recommendation domain, resulting in lesser than desired gains. Therefore, Dobermann and his team at the International

Table 16.2: Results from field verification trails of fertilizer adjustment equations of rice in inceptisols of Barddhaman

Soil series (farmer)	Soil-test values ($kg\ ha^{-1}$)			Treatment	Fertilizers dose ($kg\ ha^{-1}$)			Yield ($t\ ha^{-1}$)
	N	P_2O_5	K_2O		N	P_2O_5	K_2O	
Syamsundarpur I (T. Dutta)	245	70	162	Farmers practice	100	80	40	6.0
				General recommended dose	120	50	50	6.1
				7 $t\ ha^{-1}$ target	104	61	70	6.7
				8 $t\ ha^{-1}$ target	126	73	84	7.2
				Control	0	0	0	2.5

Source: Bera *et al.* (2006)

Rice Research Institute (IRRI) together with National Agricultural Research and Extension Systems (NARES) partners in Asia launched a research project in 1997 to develop site-specific nutrient-management (SSNM) technology for intensive rice systems (Dobermann *et al.*, 2002; Johnston *et al.*, 2009). The SSNM has been discussed in detail by them. According to Shinde *et al.* (2013), SSNM now involves: *1*, Use of nutrient omission technique to determine exact nutrient requirement; *2*, NPK doses specific to a field; *3*, Effective use of indigenous nutrients in a field; *4*, Use of effective nutrient-management techniques; *5*, use of correct amounts of necessary micronutrients.

Table 16.3. Mean and range of rice yield ($t\ ha^{-1}$) under different nutrient management practices

Year		SSNM	Recommended dose of fertilizer	Farmers' practice	LSD (P=0.05)
2004–05	Mean	8.5	7.9	7.1	0.61
	Range	7.9–9.2	7.3–8.2	6.7–7.6	
2005–06	Mean	8.1	7.1	6.3	0.55
	Range	6.7–9.5	5.6–8.5	5.1–7.2	

Source: Biradar *et al.* (2011)

Timsina, J., Jat, M.L. and Majumdar, K. (2010). Nutrient management research priorities in rice–maize systems of South Asia. *Better Crops–South Asia*: 1–6.

In India, a number of studies have been carried out on SSNM in several crops and cropping systems (Biradar *et al.*, 2006, 2011; Khurana *et al.*, 2008; Timsina *et al.*, 2010). Some data showing the advantage of SSNM in rice in Tungbhadra region of Karnataka are given in Table 16.3.

Nutrient Expert®

An easy-to-use, interactive computer-based decision tool Nutrient Expert® has been developed by the International Plant Nutrition Institute (IPNI)

(Pampolino, 2012) for making the fertilizer recommendations and its South Asia version for maize and wheat can be downloaded from http://software.ipni.net/ (Tandon, 2014). This tool takes care of crop yield, profitability and mitigation of GHG emission from the fields and in validation trials across India has shown 80–90% success rate (Sapkota *et al.*, 2014). Data from a field experiment by Mondal *et al.* (2018) on wheat in West Bengal are presented in Table 16.4, which show that fertilizer recommendations based on Nutrient Expert® gave 11% higher yield, significantly higher agronomic efficiency of N (AEn) and significantly higher B : C ratio. Also Nutrient Expert® based fertilizer recommendation had more balanced NPK application and slightly more ecofriendly reduced N rate.

Table 16.4: Relative efficiency of recommended dose of fertilizer and Nutrient Expert® in wheat in West Bengal

Treatment	Grain yield (t ha^{-1})	AEn (kg grain kg^{-1} N)	Benefit : cost (B : C) ratio
RDF	3.44	13.06	1.45
Nutrient Expert®	3.83	16.74	1.57
LSD (P=0.05)	0.48	1.2	0.11

RDF, Recommended NPK (150-26.3-33.3 kg ha^{-1}): Applied 25 kg ha^{-1} N, full P and half K using NPK mixture as basal. The remaining N was applied in 2 equal splits – at first and second irrigation and the remaining K at the second irrigation. Top-dressing just after irrigation.

Nutrient Expert® based recommendation (140-32.9-65 kg ha^{-1}): Applied 25 kg ha^{-1} N, full P and half K using NPK mixture as basal. The remaining N was applied in 2 equal splits–at first and second irrigation and the remaining half K at second irrigation. Top-dressing just before irrigation.

Source: Mondal *et al.* (2018).

Soil Test/ LCC for Top Dressing

Most fertilizer recommendations ask the farmers only a part of total N at sowing and the rest is to be top-dressed. Farmers would like to know the amount of N to be top-dressed. In some states in the USA, sampling 0–30 cm surface soil layer just before top-dressing and analyzing it for nitrate-N is used for advising N application rates (Meisinger *et al.*, 1992; Fox *et al.*, 1992); this test is known as pre-side dress soil nitrate test (PSNT). This would not be possible under Indian conditions, because time between soil sampling and N application is too small. A better alternative is to use the leaf-colour chart or SPAD meter for measuring chlorophyll concentration in leaf and decide about top- dressing amounts and schedule (Varinderpal-Singh, 2010). Some data on the use of LCC on N scheduling are given in Table 16.5, which show that same yield of rice could be obtained with 90 kg N ha^{-1} applied in 3 applications using LCC as compared to recommended practice of applying 120 kg N ha^{-1} in 3 split applications at transplanting, early tillering and panicle initiation. Application of N using LCC also significantly increased agronomic efficiency (AEn) of N. With the same level of N application (120 kg N ha^{-1}) in 4 applications the yield could be significantly increase by 0.5 t ha^{-1}.

Table. 16.5. Effect of real time nitrogen management using Leaf colour chart (LCC) in rice

Treatment	No. of applications	kg N ha^{-1} applied	Grain (t ha^{-1})	AEn (kg grain kg N^{-1})
Control	0	0	4.10	–
Recommended dose	3	120	7.10	25.0
LCCs (30 kg N ha^{-1})	4	120	7.60	29.2
LSD (P=0.05)			0.34	2.4

AEn, Agronomic efficiency of N; *Source*: Bhat *et al.* (2015)

PLANT TISSUE TESTING

Plant tissue testing is probably the best way to know the status of nutrient in plants, but it requires a lot of basic research and then a fast mechanism of collecting samples, dispatch to the laboratory for analysis and then communicating the recommendation to the farmer. As regards research, deficiency, sufficiency and toxicity (in case of micronutrients B, Mn, Cu and others if necessary) levels for a specific stage and plant part of each crop have to be determined through greenhouse or field experiments. It will take a couple of years for a research institution/ laboratory in an area. It becomes much more difficult in India, because small-holder farmers in many areas grow a wide variety of crops in a season and critical deficiency levels have to be determined for each crop. Even after this has been done, collection of samples, their drying and dispatch to the laboratory for analysis, doing the analysis and communicating the recommendation in a short span of time to permit the farmer to apply the needed amount of nutrient in time is too difficult a task in India. Plant tissue testing is nevertheless worth trying for fruit crops, where the trees are already established and the numbers of planters are fewer and crop could be just one over a large area.

STANDARD CALCULATIONS FOR DETERMINING OPTIMUM FERTILIZER DOSE

General formulae for studies on response to fertilizers (Fig. 16.3) and calculating optimum dose are given below (FAO, 1970).

$$y = a + bx + cx^2$$

where y is the yield, x is the rate of fertiizer applied in units of 1, 2, 3, etc., a is the control yield at x = 0, b is the slope at x = 0, c is the measure of downpull, if C is negative

The above factors are calculated from the values given in Fig.16.3.

a = C

b = (4A–B–3C) / 2

c = (B + C–2A)/2

The maximum dose is calculated by the exprssion $x_{max} = - b/2c$; and y_{max} is also calculated by the expression $y_{max} = a + ½ bx_{max}$

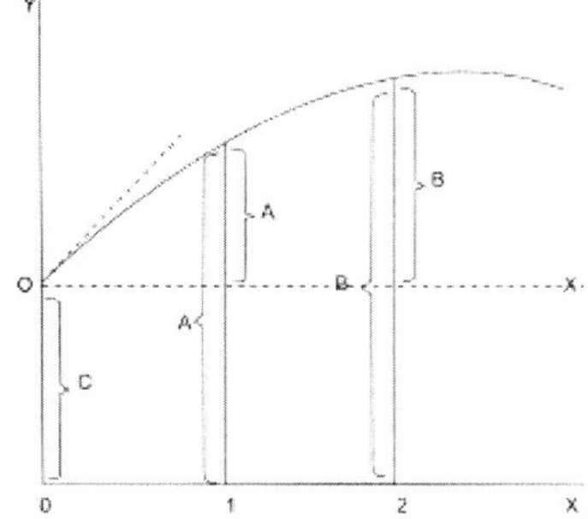

Fig. 16.3: Response curve to fertilizer (units)

SUMMARY

1. Most fertilizer recommendations in India are based on grading the soils under low, medium and high in available soil test for a nutrient.
2. In general, experience on crop response to a nutrient is utilized for making out the recommendations.
3. Some recommendations are also made on targeted yield approach.
4. Nutrient Expert® has also proved quite satisfactory in evaluation trials.
5. Plant tissue testing should first be tried on fruit crops.

REFERENCES

Bera, R., Seal, A., Bhattacharyya, P., Das, T.H., Sarkar, D. and Kangjoo, K. 2006. Targeted yield concept and a framework of fertilizer recommendation in irrigated rice domains of subtropical India. *Journal of Zhejiang University Science* B7(12): 963–968.

Bhat, T.A., Kotru, K.N., Singh, K.N., Dar, S.B. and Ram, H. 2015. Real-time nitrogen management using leaf-colour chart in rice (*Oryza sativa*) genotypes. *Indian Journal of Agronomy* **60**(1): 70–75.

Biradar, D.P., Aladakatti, Y.R. and Basavanneppa, M.A. 2011. On-farm evaluation of site–specific nutrient management in irrigated rice. *International Rice Research Notes* **36**: 0117–4185.

Biradar, D.P., Aladakatti, Y.R., Rao, T.N. and Tiwari, K.N. 2006. Site-specific nutrient management for maximization of crop yields in northern Karnataka. *Better Crops* **90**(3): 33–35.

Cate, R.B. (Jr) and Nelson, L.A. 1965. A rapid method for correlation of soil test analysis with plant response data. *North Carolina Agricultural Experiment Station, International soil Testing Series Bulletin No.* 1.

Dobermann, A., Witt, C., Dawe, D., *et al.* 2002. Site-specific nutrient management for intensive rice cropping systems in Asia. *Field Crops Research* **74**: 37–66.

Douglas, B. 2011. Chapter 14, Interpretation of Soil Testing Results. (In) *Recommended Soil Testing Procedures for the Northeastern United States, North Eastern Regional Cooperative Bulletin No.* 493. pp: 110–116.

FAO. 1970. Soil fertility investigations on farmers' fields. FAO Soils Bulletin No. 11, Food and Agriculture Organization of the United Nations, Rome.

Fox, R.H., Meisinger, J.J., Sims, T.J. and Piekielek, W.P. 1992. Predicting nitrogen fertilizer needs for corn in humid regions: advances in the Mid-Atlantic estates. (In) *Predicting Nitrogen Fertilizer Needs for Corn in Humid Region.* Bock B.R. and Kelly K.R. (Eds). National Fertilizer and Environmental Research Center, Muscle Shoals, AL, the USA, pp. 43–56.

Haokip, I.C. 2014. *Soil phosphorus dynamics in maize–wheat cropping system as influenced by phosphorus fertilization and microbial inoculants.* M.Sc. Thesis 2014, Division of Soil Science and Agricultural Chemistry, ICAR-Indian Agricultural Research Institute, New Delhi, India.

Jadhav, T.A. 2014. *Effect of different combinations of phosphorus and zinc level and their mode of application in aerobic rice–wheat cropping system.* Ph.D. Thesis, Division of Agronomy, ICAR-Indian Agricultural Research Institute, New Delhi, India.

Johnston, A.M., Khurana, H.S., Majumdar, K. and Satyanarayana, T. 2009. Site–specific nutrient management: concept, current research, and future challenges in Indian agriculture. *Journal of the Indian Society of Soil Science* **57**(1): 1–10.

Khurana, H.S., Bijay-Singh., Dobermann, A., Phillips, S.B., Sidhu, A.S. and Yadvinder-Singh. 2008. Site-specific nutrient management performance in a rice–wheat cropping system. *Better Crops* **92**(4): 26–28.

Meisinger, J.J., Bandel, V.A., Angle, J.S., O'Keefe, B.E. and Reynolds, C.M. 1992. Presidedness soil nitrate test evaluation in Maryland. *Soil Science Society of America Journal* **56**: 1,527–1,532.

Mondal, T., Mitra, B. and Das, S. 2018. Precision nutrient management in wheat (*Triticum aestivum*) using Nutrient Expert®: Growth phenology, yield, nitrogen-use efficiency and profitability under eastern sub-Himalayan plains. *Indian Journal of Agronomy* **63**(2): 174–180.

Muhr, G.R. 1965. *Soil Testing in India.* United States Agency for International Development, New Delhi.

Paramesh, V. 2014. Integrated phosphorus management and zinc fortification in maize–wheat cropping system. PhD Thesis, Division of Agronomy, ICAR-Indian Agricultural Research Institute, New Delhi, India.

Pampolino, M.F., Witt, C., Pasuquin, J.M., Johnston, A. and Fisher, M.J. 2012. Development approach and evaluation of the Nutrient Expert® software for nutrient management in cereal crops. *Computers and Electronics in Agriculture* **88**: 103–110.

Ramamoorthy, B., Narasimham, R.L. and Dinesh, R.S. 1967. Fertilizer application for specific yield target of Sonara-64 wheat. *Indian Farming* **17**: 43–45.

Rao, S. and Srivastava, S. 2000. Soil test based fertilizer use – a must for sustainable agriculture. *Fertiliser News* **45**: 25–38.

Sapkota, T.B., Majumdar, K., Jat, M.L., Kumar, A., Bishnoi, D.K., McDonald, A.J. and Pampolino, M. 2014. Precision nutrient management in conservation agriculture based wheat production of northwest India: Profitability, nutrient use efficiency and environmental footprints. *Field Crops Research* **155**: 233–244.

Shinde, V.B., Wandre, S. and Karetha, K.M. 2013. Site-specific nutrient management (SSNM). *Agrobios Newsletter* **12**(1): 40–51.

Tandon, H.L.S. 2014. History of soil fertility. (In) *Textbook of Plant Nutrient Management,* pp. 1–22. Prasad, R., Kumar, D., Rana, D.S., Shivay, Y.S. and Tewatia, R.K. (Eds.), Indian Society of Agronomy, New Delhi.

Timsina, J., Jat, M.L. and Majumdar, K. 2010. Nutrient management research priorities in rice–maize systems of south Asia. *Better Crops–South Asia*: 1–6.

Varinderpal-Singh, Bijay-Singh, Yadvinder-Singh., Thind, H.S. and Gupta, R.K. 2010. Need-based nitrogen management using the chlorophyll meter and leaf color chart in South Asia: A review. *Nutrient Cycling in Agroecosytems* **88**(3): 361–380.

QUESTIONS

1. Briefly discuss the followings

(a) Nutrient Expert®
(b) Graphic method of determining the critical level of a nutrient.
(c) Targeted yield concept.
(d) Problems involved in using 'Plant Tissue Testing' for field crop production.
(e) LCC for making fertilizer recommendations.

2. Match the followings

(a)	Agronomic efficiency	(i)	kg nutrient t^{-1} grain or kg nutrient q^{-1} grain
(b)	Plant Tissue Testing	(ii)	Ramamoorthy
(c)	Cate and Nelson	(iii)	kg grain kg N^{-1}
(d)	Nutrient requirement	(iv)	PSNM
(e)	Top dressing of a nutrient	(v)	Graphic method of determining critical deficiency level
(f)	Targeted yield concept	(vi)	Fruit crops

Chapter 17

Integrated Nutrient Management

The introduction of high-yielding varieties of wheat and rice, the development of intensive cereal–cereal rotations (rice–wheat, rice–rice, hybrid maize–wheat), application of high doses of N (above 100 kg N ha^{-1} to each crop) without adequate P and K application, increased use of high-analysis fertilizer (urea and diammonium phosphate in place of ammonium sulphate and superphosphate) and ignoring the application of organic manures led to heavy depletion of P, K, S, Zn and other plant nutrients, which led to wide-spread deficiencies of P, S and Zn in Indian soils. These deficiencies of nutrients led to focus on integrated nutrient management.

Integrated nutrient management (INM) or integrated plant nutrient supply system (IPNS) is an approach, which adapts plant nutrition to a specific farming system and particular yield targets, the resource base, the available plant nutrient source and socio-economic background (Dudal and Roy, 1995). Thus, INM demands a holistic approach to nutrient management for agricultural production. For a true INM programme, regular monitoring of all incoming nutrients including those from irrigation water and outgoing nutrients including leaching losses has to be done. However, for all practical purposes, INM involves a judicious combined use of chemical fertilizers, organic manures, crop residues and biofertilizers and inclusion of legumes in a cropping system either as a grain/ green vegetable/ fodder crop or as a green-manure. The role of organic manures, crop residues and legumes in nutrient management has already been discussed in previous chapters.

INTEGRATED NUTRIENT MANAGEMENT AND CROP PRODUCTION

The advantage of the combined use of organic manures and chemical fertilizer is well brought out by experiments under long-term experiments of All India Coordinated Research Project (AICRP) on Cropping Systems (Table 17.1). The data in Table 17.1 clearly reveal that over a long period of time, application of 50% N through farmyard manure (FYM) or crop residue or green-manure in the rainy season (*kharif*) followed by 100% (RDF) through fertilizer resulted in higher production in rice–rice cropping system.

However, Singh *et al.* (2018) failed to obtain any advantage of INM using FYM and vermicompost in pearl millet at Jammu over a period of 3 years. Thus, benefits of INM on crop production could vary from place to place.

INTEGRATED NUTRIENT MANAGEMENT AND NUTRIENT-USE EFFICIENCY

Efficiency of applied N, P and K, in general increases, when FYM is applied along with these nutrients (Table 17.2). Part of this could be owing to nutrients added by FYM, which is not accounted in these experiments.

INTEGRATED NUTRIENT MANAGEMENT AND SOIL PROPERTIES

In addition to sustaining production, INM also leads to better soil quality by improving chemical, physical, physico-chemical and biological properties.

Table 17.1: Effect of 50% substitution on nitrogen (N) by farmyard manure (FYM)/ crop residue/ green manuring on the productivity (t ha^{-1}) of rice–rice cropping system at four research centres (results averaged over 15–21 years)

Treatment	Bhubaneshwar (Odisha)	Maruteru (Andhra Pradesh)	Karjat (Maharashtra)	Karamana (Kerala)
100% RDF through fertilizer in summer (*kharif*) and winter (*rabi*)	9.0	10.7	8.2	7.6
50% N through FYM in *kharif* + 50% RDF as fertilizer in *kharif* + 100% RDF in *rabi* as fertilizer	9.8	10.5	8.2	7.6
50% N through crop residue in *kharif* + 50% RDF as fertilizer in *kharif* + 100% RDF in *rabi* as fertilizer	9.4	10.4	7.9	7.9
50% N through green manure/ green leaf manure in *kharif* + 50% RDF as fertilizer in *kharif* + 100% RDF in *rabi* as fertilizer	9.7	10.7	8.5	8.5

RSF, Recommended dose of fertilizers

Source: Gill *et al.* (2008)

Soil *p*H

In AICRP on Long-Term Fertilizer Experiments (LTFE) at Ranchi, continuous application of inorganic fertilizers over a period of 40 years reduced the soil *p*H from 5.4 (control) to 3.9–4.2, while application of FYM along with NPK increased it to 6.3–6.5 (Sarkar and Singh, 2002). Exchangeable Ca increased from 3.05 in the control plots to 8.25 cmol (+) kg^{-1} in plots receiving FYM.

Table 17.2: Nutrient-use efficiency (%) of nitrogen, phosphorus and potassium at four centres under Long-term Fertilizer Experiments (LTFE)

Treatment	Ludhiana		Udaipur		Raipur	Pattambi
	Maize	Wheat	Maize	Wheat	Kharif rice	Rabi rice
Nitrogen						
100% N	12	26	11	23	20	10
100% NP	19	41	18	34	39	13
100% NPK	24	48	23	39	40	16
100% NPK + FYM	37	54	31	50	46	27
Phosphorus						
100% N	7	15	6	8	11	6
100% NP	12	24	10	12	21	8
100% NPK	14	27	12	15	22	11
100% NPK + FYM	22	31	16	18	25	17
Potassium						
100% N	62	124	69	137	66	25
100% NP	96	198	116	202	126	33
100% NPK	119	230	146	236	129	42
100% NPK + FYM	186	257	194	297	148	69

Source: Singh *et al.* (2019); Maize in *kharif* (July–October) and wheat in *rabi* (November–April) season

Soil organic carbon (SOC) and plant nutrients

In AICRP on Long-Term Fertilizer Experiments (LTFE), application of FYM with NPK increased SOC by 7.7% at Ludhiana, 49.3% at Jabalpur, 33.9% at Bengaluru, 14.2% at Ranchi and 50% at Bhubaneshwar over a period of 15–25 years (Swarup, 2002). Similarly, in long-term experiments under AICRP on Cropping Systems, application of 50% N as FYM + 50% RDF as chemical fertilizer to rice followed by 100% RDF as chemical fertilizer to wheat for 8–9 years resulted in increase in SOC by 31% at Faizabad and 18% at Sabour (Hegde, 1998). Increase on SOC by addition of FYM in rice–wheat cropping system has been reported by Prasad and Mishra (2001). A 7-year study at Hisar (Haryana) in pearlmillet–wheat cropping sequence showed that the application of farmyard manure, poultry manure, and sugarcane filter cake alone or in combination with chemical fertilizers improved the soil organic C and total N, P, and K status (Kaur *et al.*, 2005).

Soil physical properties

Application of FYM in rice improved the bulk density and soil structure and resulted in an increase in grain yield by 25% at Umiam (Meghalaya) (Mishra and Saha, 2008). In a soybean–wheat–maize (fodder) crop rotation on Vertisol at Jabalpur, SOC content in 100% NPK and 100% NPK + FYM treatments increased by 22.5 and 56.3%, respectively, over the initial level (1.14 kg m^{-3}).

The electrical conductivity, SOC content, aggregation, water retention, microporosity and available water capacity of the soil were increased, while the bulk density was reduced significantly with the 100% NPK + FYM treatment over all other treatments (Hati *et al.*, 2007).

Soil biological properties

Application of FYM @ 10 t ha^{-1} $year^{-1}$ in sugarcane over 2 year increased the soil microbial biomass carbon (SMBC) by 21.9% and soil microbial biomass nitrogen (SMBN) by 48.8% over the control as compared to an increase of 5.6% in SMBC and 27.9% in SMBN obtained with 150 kg N + 60 kg P_2O_5 + 60 kg K_2O ha^{-1} $year^{-1}$ (Singh *et al.*, 2007). Kaur *et al.* (2005) from Hisar reported an increase in soil microbial biomass C and N in soils receiving organic manures only or with the combined application of organic manures and chemical fertilizers compared to soils receiving chemical fertilizers only. Basal and glucose induced respiration, potentially mineralizable N, and arginine ammonification were higher in soils amended with organic manures with or without chemical fertilizers, indicating that more active microflora is associated with INM. Studies from a permanent field experiment on sorghum –wheat cropping sequence continuing since 1984–85 on a swell-shrink soil (Vertic Haplustept) at Research Farm under AICRP on Integrated Farming Systems Research Unit, Akola (Maharashtra), showed that the population of bacteria (87.5×10^6 cfu), fungi (15.25×10^4 cfu) and actinomycetes (16.25×10^4 cfu) significantly increased with the application of 50% N through FYM along with 50% RDF. Some data from Pantnagar, Uttarakhand, are given in Table 17.3. The INM using FYM also exerted a profound effect SMB C (267 mg kg^{-1}), CO_2 evolution (50.3 mg 100 g^{-1} soil), dehydrogenase (36.7 µg TPF g^{-1} d^{-1}), urease (78.4 µg NH_4–N g^{-1} d^{-1}) and alkaline phosphatase activity (194 µg p-nitrophenol g^{-1} d^{-1}). Similarly, green-manuring with *Leucaena* loppings and crop residue recycling using wheat straw in conjunction with fertilizers recorded significant improvement in soil biological properties over fertilizers alone. The biological properties were drastically deteriorated where no organics were used for 27 years (Mali *et al.*, 2015).

SUMMARY

1. Continuous growing of high-yielding varieties of cereals with high-analysis fertilizers in intensive cropping systems has depleted soil fertility.
2. Use of organic manures replenishes the soil with all the plant nutrients including N, P, K, S and micronutrients.

Table 17.3: Long-term effect of fertilizer and manure on microbial population in soil (After 44 crop cycles at Pantnagar centre of All India Coordinated Research Project on Long-term Fertilizer Experiments)

Treatment	Fungi ($\times 10^4$)	Bacteria ($\times 10^6$)	Actinomycetes ($\times 10^5$)	*Azotobacter* ($\times 10^3$ cfu g^{-1})
Control	0.60	0.55	0.78	152
100% N	1.27	1.46	1.89	410
100% NP	1.66	1.68	1.00	476
100% NPK	1.48	1.80	1.71	286
100% NPK + FYM	1.99	2.32	2.16	630

Source: Singh *et al.* (2019)

3. The INM involving judicious combination of chemical fertilizers, organic manures, crop residues, green manures and biofertilizers is the only way for sustainable agriculture.
4. The INM results in improved soil physical, chemical and biological properties.

REFERENCES

Dudal, R. and Roy, R.N. 1995. *Integrated Plant Nutrient Systems*. FAO Fertilizer and Plant Nutrition Bulletin **12**: 181–198.

Gill, M.S., Gangwar, B. and Gangwar, K.S. 2008. Site-specific crop management approach or high yield realization of cereal-based cropping systems. *Indian Journal of Fertilisers* **4**(8): 31–56.

Hati, K.M., Swarup, A., Dwivedi, A.K., Misra, A.K. and Bandypadhyay, K.K. 2007. Changes in soil physical properties and organic carbon status at the top soil horizon of a Vertisol of central India after 28 years of continuous cropping, fertilization and manuring. *Agriculture, Ecosystems and Environment* **119**(2): 127–134.

Hegde, D.M. 1998. Integrated nutrient management effect on rice–wheat system productivity in sub-humid ecosystems. *Indian Journal of Agricultural Sciences* **68**: 144–148.

Kaur, K., Kapoor, K. and Gupta, A.P. 2005. Impact of organic manures with and without mineral fertilizers on soil chemical and biological properties under tropical conditions. *Journal of Plant Nutrition and Soil Science* **168**: 117–122.

Mali, D.V., Kharche, K.V., Jadhao, S.D., Jadhao, S.M., Saoji, B.V. and Age, A.B. 2015. Soil biological health under long-term fertilization in sorghum–wheat sequence on swell-shrink soils of central India. *Journal of the Indian Society of Soil Science* **63**(4): 423–428.

Mishra, V.K. and Saha, R. 2008. Soil physical behaviour and rice (*Oryza sativa*) under different sources of organics, methods of puddling and zero tillage. *Indian Journal of Agricultural Sciences* **78**: 399–404.

Prasad, R. and Mishra, B.N. 2001. Effect of addition of organic residues, farmyard manure and fertilizer nitrogen on soil fertility in rice–wheat cropping system. *Archives of Agronomy and Soil Science* **46**: 455–463.

Raychaudhuri, S.P. 1984. Development of soil productivity in ancient and mediaeval India. (In) *Soil Science in India, Indian Society of Soil Science, New Delhi, Bulletin No.* 14: 1–13.

Sarkar, D.A.K. and Singh, R.P. 2002. Importance of long-term fertilizer use for sustainable agriculture in Jharkhand. *Fertiliser News* **47**(1): 107–111.

Singh, B., Kumar, A., Abrol, V., Singh, A.P., Kumar, J. and Sharma, A. 2018. Effect of integrated plant nutrient management on pearlmillet (*Pennisetum glaucum*) productivity in rainfed subtropic Shiwalik foothills of Jammu and Kashmir. *Indian Journal of Agronomy* **63**(2): 197–200.

Singh, K.P., Srivastava, T.K., Singh, P.N. and Suman, A. 2007. Enhancing soil fertility, microbial activity and sugarcane (*Saccharum officinarum*) productivity through organics in subtropical conditions. *Indian Journal of Agricultural Sciences* **77**(2): 84–87.

Singh, M., Wanjari, R.H., Kumar, U. and Chaudhari, S.K. 2019. AICRP on Long-Term Fertilizer Experiments: Salient achievements and future directions. *Indian Journal of Fertilisers* **15**(4): 356–372.

Swarup, A. 2002. Lessons from long-term fertilizer experiments in improving fertilizer use efficiency and crop yields. *Fertiliser News* **47**(12): 59–73.

QUESTIONS

1. Briefly discuss the concept of INM.
2. What are effects of organic manures on chemical properties of soil?
3. How does INM affect soil biological properties?

4. True/ False

(a) Benefits of INM on crop production cannot vary from place to place.
(b) Soil biological properties are drastically deteriorated where no organics are used.
(c) INM may or may not improve soil physical, chemical and biological properties.
(d) The continuous application of inorganic fertilizers over a long period increases the soil *p*H while application of FYM along with NPK decreases it.
(e) The application of FYM with NPK increases SOC in rice–wheat cropping system.

Answers

(a) False (b) True (c) False (d) False (e) True

5. Fill in the blanks

(a) Continuous growing of high-yielding varieties of cereals with high-analysis fertilizers in intensive cropping systems has depleted ....
(b) The INM involves a judicious combined use of chemical fertilizers, organics manures, green manuring, crop residues and .... is the only way for sustainable agriculture.
(c) Use of .... replenishes the soil with all the plant nutrients including macro and micronutrients.

(d) In long-term application of 50% N through FYM along with 50% RDF showed a significant .... in the population of bacteria, fungi and actinomycetes in sorghum–wheat cropping system.

(e) In soybean–wheat–maize (fodder) crop rotation, the application of 100% NPK + FYM were increased EC, SOC content, aggregation etc. while the .... was reduced significantly.

Answers

(a) Soil fertility (b) Biofertilizers (c) Organic manures

(d) Increase (e) Bulk density

Chapter 18

Environmental Implication of Fertilizers

Plant nutrients are important for increased production of food for the ever-increasing world population. Of the plant nutrients, primary nutrients N, P and K are applied to fields each season in fairly large amounts. In addition to increasing crop yields, they also impact the environment. This is briefly discussed in this chapter; only main points are made here.

NITROGEN

Nitrogen is blamed the most for environmental degradation. It contributes to eutrophication of inland and sea-waters, enrichment of air with NH_3 and N_2O and even with global warming. However, it is almost impossible to grow more food without additional nitrogen. It is estimated that by 2050 about 250 million tonnes (MT) of fertilizer nitrogen may be needed to meet global increased food demands as against an estimated current consumption of 116 MT (Tilman *et al.*, 2011).

Eutrophication of inland waters

When heavy rains follow fertilizer N application, especially on sloping lands, part of applied N may be lost as surface run-off. Rice farmers in sloping lands in India are very familiar with this loss and it deters them from applying recommended N doses. Studies on N runoff trends in China during the period of 1990–2012, showed that N runoff has increased by 46% for rice paddy fields and 31% for upland areas since 1990 (Xikang *et al.*, 2017). Van Drecht *et al.* (2003) using Integrated Model to Assess the Global Environment (IMAGE) reported, that in Western Europe and Japan 70–75 kg N ha^{-1} was exported from the point of application, the suggested values for South Asia, East Asia, Canada and the USA were 30–40 kg N ha^{-1}.

Eutrophication of waters with N results in algal blooms. The term '*algal bloom*' refers to the rapid growth and accumulation of large mass of phytoplankton (cyanobacteria, algae, diatoms, dinoflagellates) in waters, which results in hypoxia (lack of oxygen) leading to the mortality of fish and other aquatic animals. Such areas in the seas are known as '*Dead Zones*' and as many as

405 have been recognized in the world including coasts of Gulf of Mexico, Chesapeake Bay, South China sea, East China sea, Arabian sea and Bay of Bengal (Diaz and Rosenberg, 2008). From India, Padmakumar *et al.* (2012) reported that during 1998–2010, eighty algal blooms were recorded, of which 31 were due to dinoflagellates, 27 by cyanobacteria and 18 by diatoms. Recently, algal blooms were blamed for a dip in fish catch from sea coast at Bombay by an Indo-US team of experts (ToI, 2018). Algal blooms lead to fish mortality, affect drinking water supply and make lakes risk prone for entertainment. Toxins produced by algal blooms also lead to illness and mortality of cattle (Hilborn and Beasley, 2015).

Ammonia enrichment of air

A major portion of N fertilizer applied to soil is lost through volatilization as gaseous NH_3 (AVL), which acts as the precursor of several atmospheric pollutants (Aneja *et al.*, 2012). Ammonia is the most reduced form of reactive nitrogen and most abundant of alkaline constituent in the atmosphere. Agriculture sector contributes more than 80% of total anthropogenic NH_3 emission to the atmosphere. Application of N fertilizer in soil contributes more than 25% of the total NH_3 emission from agriculture. The emission from soil varies from 1 to 26 kg ha^{-1} depending on soil type, crop and fertilizer material (Adhya *et al.*, 2007). Some other studies have shown that ammonia volatilization loss is about 20–30% of N applied through fertilizer and manure (Pathak *et al.*, 2006). Neem-cake/ neem oil microemulsion coating of urea can reduce AVL (Chauhan and Misra, 1989; Blaise and Prasad, 1995: Prasad *et al.*, 1999).

Nitrate enrichment of groundwater

Leaching of N in the NO_3 form beyond soil profile is one of the major loss mechanisms that can be as high as 80% depending on the soil properties and water and nutrient-management practices. It is particularly a major sink for surplus N in agro-ecosystems on well-drained porous soils. A lot of attention has already been paid the world over, particularly in the developed countries on this phenomenon. In developing countries like India, its importance is being felt recently because of sporadic reports of NO_3 enrichment in groundwater in a few agriculturally intensive areas in Punjab, Haryana, Delhi, Maharashtra, and Andhra Pradesh. They have attributed the high NO_3-N in groundwater to high nitrogenous fertilizer use. Bharadwaj *et al.* (2012) reported that, 28% of shallow well waters in Ludhiana district contained 11–25 mg NO_3-N L^{-1}, which is above the EPA safe limit. Vinod *et al.* (2015) reported that, in Karnataka, the average nitrate-N leached from irrigated areas varied from 51.2–74.9 kg N ha^{-1}. But there were several exceptions, for example, in West Bengal; the average NO_3 level was low in spite of high dosage of N fertilizer use (Kundu *et al.*, 2008). Sporadic reports of high NO_3-N contamination in groundwater from different parts of India may be related more to a sewage-sewerage and dumping of organic wastes rather than agricultural N fertilization barring a few sites where the soil is highly coarse textured and hence permeable. Using crop-simulation model Pathak *et al.* (2006) calculated that, N losses due to leaching from rice–wheat cropping systems of the Indo-Gangetic plains could account for 10% of applied N. Nitrate in groundwater is responsible for methaemoglobinemia or blue baby syndrome (Knobeloch *et al.*, 2000). The safe limit of nitrate in drinking water as per Environmental Protection Authority (EPA) is 10 mg NO_3-N L^{-1}.

A recent study in Punjab investigated the concentration of NO_3 in groundwater of 14 intensively cultivated districts of Malwa Punjab (Ahada and Suthar, 2018). The sampling of 76 sites revealed the concentration were in ranges of 38.45–198.05 mg L^{-1} and over 92% sites showed the high level of it than the safe limits for drinking water, which is 10 mg NO_3–N L^{-1} (45 mg NO_3 L^{-1}).

N_2O, ozone layer depletion and global warming

Most of the applied N is finally oxidized to nitrate, a part of which is reduced to nitrite that produces N_2O. The process is known as denitrification. Denitrification, a sequential reduction of NO_3^- to N_2O driven by denitrifying bacteria under anaerobic conditions and is controlled by soil moisture, redox potential (Eh), temperature, *p*H and substrate (NO_3, NO_2, NO and N_2O) concentrations. Denitrification is usually not explicitly accounted for N budgets on the national level, but estimated indirectly from N balances. Yet, it is generally accepted that denitrification is a major sink for the surplus N in agro-ecosystems. On an average, denitrification losses reported in India ranged from 10 to 30 and from 5 to 10 kg N ha^{-1} in rice and wheat, respectively (Aulakh *et al.*, 2001).

Nitrogen loses due to denitrification are likely to be most under alternate flooding and drying as obtained under aerobic rice systems (Prasad, 2011). It is estimated that globally, about 38% of nitrous oxide in the atmosphere is of anthropogenic origin out of which agriculture may account for about 67 to 84% and out of the total contribution of agriculture, fertilizer N may contribute about one-fourth (6.36% of total anthropogenic contribution) (IPCC, 2007; Smith, 2010). In recent years, fertilizer N has been a major contributor to atmospheric N_2O (Synder *et al.*, 2014). Neem-coating of urea can reduce N_2O emission from soil. Majumdar *et al.* (2000, 2002) reported that, N_2O emission in rice–wheat cropping system was more in wheat (750–1430 g N_2O-N ha^{-1} over a period of 90 days) than in rice (34.3–59.8 g ha^{-1} over a period of 70 days). They also reported that in rice, neem cake-coated urea (NCCU) reduced N_2O emission by 11.2% as compared to 4.8% with Nimin (a commercial product containing neem bitterns), while in the case of wheat, the reduction in N_2O emission was 63% with nimin and only 9% with NCCU. The N_2O is responsible for ozone-layer depletion (Ravishankara *et al.*, 2009). Nitrous oxide (N_2O) has also a global warming power (GWP) 265–298 times that of CO_2 for a 100-year timescale. In the USA, agriculture was responsible for 7% of total greenhouse gases (GHG) of the country and about 80% it was due to N_2O emission from fertilizer nitrogen (EPA, 2009).

PHOSPHORUS

Phosphorus has received attention from two points. Firstly from the viewpoint of eutrophication of water, and secondly from the viewpoint of cadmium (Cd) accumulation in soil and its movement to the food chain, which is not an environmental problem.

Phosphorus does not leach, but is lost in surface runoff. The global agricultural P budget indicated that the average annual P accumulation in agricultural areas of the world was 8 Tg $year^{-1}$ from 1958 to 1998; developed countries contributing about 60% (Benett *et al.*, 2001). Phosphorus availability is considered critical for algal growth, since most algae can fix atmospheric N (Schindler *et al.*, 2008). Redfield (1934) had suggested a universal N : P ratio of 16 for good algal growth. Correll (1999), however, pointed out that while P is the limiting nutrient in inland waters, N is the limiting nutrient in marine waters. This is because nitrogen fixation is not observed at salinity levels greater than 10 (g kg^{-1} water), while the mean salinity in marine waters is ~35 (Paerl and Fulton III, 2006). The safe limit for not having algal blooms is ~ 0.1 mg phosphate L^{-1}. In India, phosphate concentration in Diggi pond in Aligarh in Uttar Pradesh (Khan *et al.*, 2014), and lake, groundwater and channel waters in Mysuru district, Karnataka were 1.87 to 6.79 mg L^{-1} (Divya and Belagali, 2012). Some data on nitrate and phosphate concentration in the Ganges River at three points are in Table 18.1. However, it may be pointed out that fertilizer is not the only source of phosphates in the inland waters. Phosphates also enter water ways from human and animal waste, detergents used for washing clothes and industrial effluents.

POTASSIUM

No environmental problems are reported for potassium fertilizers.

SUMMARY

1. Both N and P fertilizers are responsible for eutrophication of inland and marine waters.

Table 18.1: Nitrate and phosphate concentration in Ganges River water at three locations

Location	Quality parameter	Summer	Monsoon season	Winter
Kanpur	*p*H	8.80	8.20	8.50
	Nitrate (mg L^{-1})	1.70	0.45	0.94
	Phosphate (mg L^{-1})	1.58	0.66	0.82
	Nitrate : phosphate ratio[1]	1.07	0.68	1.14
Allahabad	*p*H	8.20	7.60	7.90
	Nitrate (mg L^{-1})	1.50	1.40	0.23
	Phosphate (mg L^{-1})	1.50	0.63	0.78
	Nitrate : phosphate ratio	1.00	2.22	0.29
Varanasi	*p*H	8.20	7.70	8.00
	Nitrate (mg L^{-1})	2.60	2.40	2.10
	Phosphate (mg L^{-1})	1.42	1.11	1.37
	Nitrate : phosphate ratio	1.83	2.16	1.53

Source: Tiwari *et al.* (2016); 1. Calculated by the authors

2. Phosphorus could be a limiting nutrient for algal growth in inland waters, while N could be a limiting nutrient in marine waters.
3. Algal blooms in waters are responsible for the death of fish and other aquatic life.
4. Nitrogen leaching can enrich underground waters with nitrates beyond limits fixed for drinking waters.
5. The safe limit for nitrates is 10 mg L^{-1} of NO_3-N or 45 mg nitrates L^{-1}.
6. Nitrate fertilizers result in the production of nitrous oxide gas (N_2O), which can damage the ozone layer and can also cause global warming.

REFERENCES

Adhya, T.K., Pathak, H. and Chhabra, A. 2007. N-Fertilizers and gaseous – N emission from rice based cropping systems. (In) *Agricultural Nitrogen Use and its Environmental Implications*, pp. 459–475. Abrol, Y.P., Raghuram, N. and Sachdev, M.S. (Eds.), New Delhi. IK International Publishing House Pvt Ltd, New Delhi.

Ahada, C.P.S. and Suthar, S. 2018. Ground water nitrate contamination and associated human health risk assessment in southern districts of Punjab, India. *Environmental Science and Pollution Research* **25**: 25,336–25,347.

Aneja, V.P., Schlesinger, W.H., Erisman, J.W., Behra, S.N., Sharma, M. and Battye, W. 2012. Reactive nitrogen emission from crop and livestock farming in India. *Atmosphere Environment* **47**: 92–103.

Aulakh, M.S., Khera, T.S., Doran, J.W. and Bronson, K.F. 2001. Denitrification, N_2O and CO_2 fluxes in rice–wheat cropping system as affected by crop residues, fertilizer N and legume green manure. *Biology and Fertility of Soils* **34**: 375–389.

Benett, E.M., Carpenter, S.R. and Caracao, N, F. 2001. Human impact on erodible phosphorus and eutrophication: A global perspective. *BioScience* **15**: 227–234.

Bharadwaj, A., Garg, S., Sondhi, S.K. and Taneja, D.S. 2012. Nitrate contamination of shallow aquifer ground water in the central districts of Punjab, India. *Journal of Environmental Science and Engineering* **54**(1): 90–97.

Blaise, D. and Prasad, R. 1995. Effect of blending urea with pyrite or coating with polymer on ammonia volatilization losses from surface applied urea. *Biology and Fertility of Soils* **20**(1): 83–95.

Chauhan, H.S. and Misra, B.1989. Ammonia volatilization from a flooded rice field with amended urea materials. *Fertilizer Research* **19**(1): 57–63.

Correll, D.L. 1999. Phosphorus: A rate limiting nutrient in surface waters. *Poultry Science* **78**: 674–682.

Diaz, R.J. and Rosenberg, R. 2008. Spreading dead zones and consequences for marine ecosystems. *Science* **321**(5891): 926–929.

Divya, J. and Belgali, S.L. 2012. Impact of chemical fertilizers on water quality in selected agricultural areas of Mysuru district, Karnataka, India. *International Journal of Environmental Sciences* **2**(3): 1449–1458.

EPA. 2009. United States Environmental Protection Agency. Inventory of US greenhouse gas emission and sinks: 1990–2007. EPA 430-R-09-004.

Hilborn, E.D. and Beasley, V.R. 2015. On health and cyanobacteria in freshwater systems: animal illnesses and deaths are sentinel events for human health risks. *Toxins (Basel)* **7**(4): 1,374–1,395.

IPCC. 2007. Climate Change, 2007. *Mitigation of Climate Change*. Cambridge University Press, New York, the UK.

Khan, F.A., Naushin, F., Masoodi, A., Ifran, M., Hashmi, F. and Ansari, A.A. 2014. Eutrophication: global scenario and local threat to dynamics of aquatic system. (In) *Eutrophication: Causes, Consequences and Control*, vol 2. Ansari A.A. and Gill S.S. (Eds). Springer (http://www.springer.com/978-94-007-7813-9).

Knobeloch, L., Saha, B., Hogan, A., Pestle, J. and Anderson, H. 2000. Blue babies and nitrate contaminated well water. *Environmental Health Perspective* **108**(7): 675–678.

Kundu, M.C., Mandal, B. and Sarkar D. 2008. Assessment of the potential hazards of nitrate contamination in surface and groundwater in a heavily fertilized and intensively cultivated district of India. *Environment Monitoring and Assessment* **146**: 183–189.

Majumdar, D., Kumar, S., Pathak, H., Jain, M.C. and Kumar, U. 2000. Reducing nitrous oxide emission from an irrigated rice field of North India with nitrification inhibitors. *Agriculture, Ecosystems and Environment* **81**: 163–169.

Majumdar, D., Pathak, H., Kumar, S. and Jain, M.C. 2002. Nitrous oxide insecticide emission from a sandy loam Inceptisol under irrigated wheat in India as influenced by different nitrification inhibitors. *Agriculture, Ecosystem and Environment* **91**: 283–293.

Padmakumar, K.B., Menon, R. and Sanjeevan, V.N. 2012. Is occurrence of harmful algae blooms in the exclusive economic zones of India on the rise? *International Journal of Oceanography* vol. 2012 Article ID 263946,7 pages (htt://dx.doi.org/10.1155/2012/263946).

Paerl, H.W. and Fulton III, R.S. 2006. Ecology of harmful cyanobacteria (In) *Ecology of Harmful Marine Algae*, pp: 95–109. Graneli E. and Turner J. (Eds), Springer, Berlin.

Pathak, H., Li, C., Wassermann, R. and Ladha, J.K. 2006. Simulation of nitrogen balance in the rice–wheat systems of the Indo-Gangetic plains. *Soil Science Society of America Journal* **70**: 1,612–1,622.

Prasad, R. 2011. Aerobic rice systems. *Advances in Agronomy* **111**: 207–247.

Prasad, R., Singh, S., Saxena, V.S. and Devakumar, C. 1999. Coating of prilled urea with neem (*Azadirachta indica* Juss.) oil for efficient nitrogen use in rice. *Naturwissenscheften* **86**: 538–539.

Ravishankara, A.R., Daniel, J.S. and Portmann, R.W. 2009. Nitrous oxide (N_2O): The dominant ozone-depleting substance emitted in the 21st Century. *Science* **326**: 123. (DOI.10.1126/science.1176985).

Redfield, A.C. 1934. On the proportion of organic derivatives in sea water in relation to the composition of phytoplankton. (In) *James Johnston Memorial Volume*. James J., Jellicoe D.R. (Eds.), University of Liverpool, Liverpool, the U.K.

Schindler, D.W., Hecky, R.E., Findlay, D.L., *et al.* 2008. Eutrophication of lakes cannot be controlled by reducing nitrogen inputs: Results of a 37 year whole ecosystem experiment. *Proceedings National Academy of Sciences, USA* **105**: 11,254–11,258.

Smith, K. (Ed.). 2010. *Nitrous Oxide and Climate Change*. Earthscan Ltd, London, the UK.

Synder, C.S., Davidson, E.A., Smith, P. and Venterea, R.T. 2014. Agriculture: sustainable crop and animal production to help mitigate nitrous oxide emissions. *Current Opinion on Environment and Sustainability* **9–10**: 46–54.

Tilman, D., Blazer, C., Hill, J. and Befort, B.L. 2011. Global food demand and the sustainable intensification of agriculture. *Proceedings of the National Academy of Sciences, USA* **108**: 20,260–20,264.

Tiwari, A., Dwivedi, A.C. and Mayank, P. 2016. Time scale changes in the water quality of the Ganges River, India and estimates of sustainability for exotic and hardy fishes. *Hydrology Current Research* **7**: 254 (doi: 10.4172/2157-7587-1000254).

ToI. 2018. Algae blooms leading to fish death of Mumbai. *Times of India, Mumbai*, 4 May, 2018.

Van Drecht, G., Bouwman, A.F., Knoop, J.M., Beusen, A.H.W. and Meinardi, C.R. 2003. Global modeling of the fate of nitrogen from point and non-point sources in soils, ground water and surface water. *Global Biogeochemical Cycles* **17**(4): 1115. doi:10.1029/2003GB002060.

Vinod, P.N., Chandramouli, P.N. and Koch, M. 2015. Estimates of nitrate leaching in ground water in agriculturally used urea in the state, Karnataka, India using existing models and GIS. *International Conference on Water Resource, Coastal and Ocean Engineering*, March 12–14, 2015, National Institute of Technology, Surratha, Karnataka.

Xikang, H., Zhan, X., Zhou, F., *et al.* 2017. Detection and attribution of nitrogen runoff trend in China's croplands. *Environmental Pollution* **234**: 270–272.

QUESTIONS

1. Fill up the blanks

(a) Nitrous oxide destroys .... layer.
(b) Nitrous oxide's GWP is .... times that of CO_2.
(c) Nitrous oxide is produced in soil by the process known as ....

2. Answer True (T) or False (F)

(a) Phosphate is the limiting nutrient for algal growth in marine waters.
(b) Nitrates in excess of 45 mg L^{-1} in drinking water can cause blue baby syndrome.
(c) Accumulation of nutrients in marine waters is termed as eutophication.
(d) Fertilizer nitrogen has the sole responsibility for eutrophication of inland waters.

Subject Index

G

I

K

L

M

N

Author Index

A

B

C

D

E

F

G

H

I

J

K

L

M

N

O

P

Q

R

S

T

U

V

W

X

Y

Z

Colour Plates

Chapter 1: Soils as the Medium for Crop Production

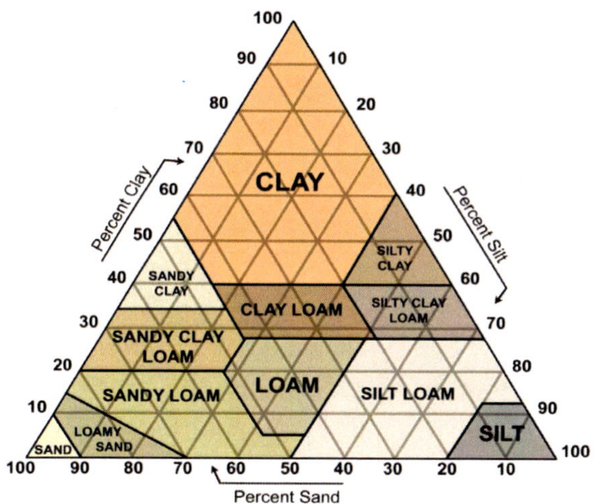

Fig.1.1: Chart for determining soil textural class (Soil Survey Staff, USDA, 1999)

Chapter 3: Essential and Beneficial Plant Nutrients

Fig. 3.1: Deficiency symptoms of some nutrients in plants.
(a) N deficiency in maize (causing dying lower leaves); (b) P deficiency in sugarcane (resulting purpling of leaves, poor tillering); (c) K deficiency in wheat (dying top leaves); (d) Zn deficiency in wheat resulting interveinal chlorosis in young (*left*) leaves, yellow-brown spots (*right*); (e) Yellowing of young leaves with dead margins due to S deficiency in sugarcane; (f) B deficiency in pineapple (Fruit necrosis which begins with the browning of inner most part of the mesocarp).
(*Source:* Vikaspedia–A Government of India portal)

Chapter 8: Soil and Fertilizer Phosphorus

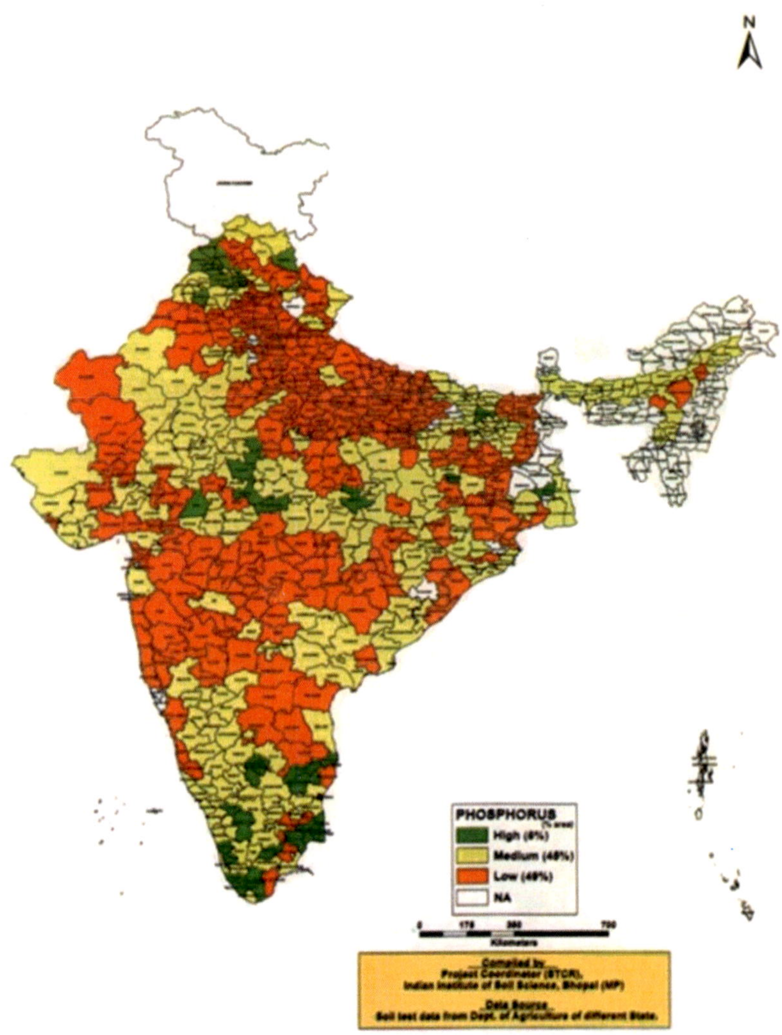

Fig. 8.1: Available P status of Indian soils. (with permission from Dr Pradip Dey, Project Coordinator, STCR, Indian Institute of Soil Science, Bhopal)

Chapter 10: Secondary Nutrients

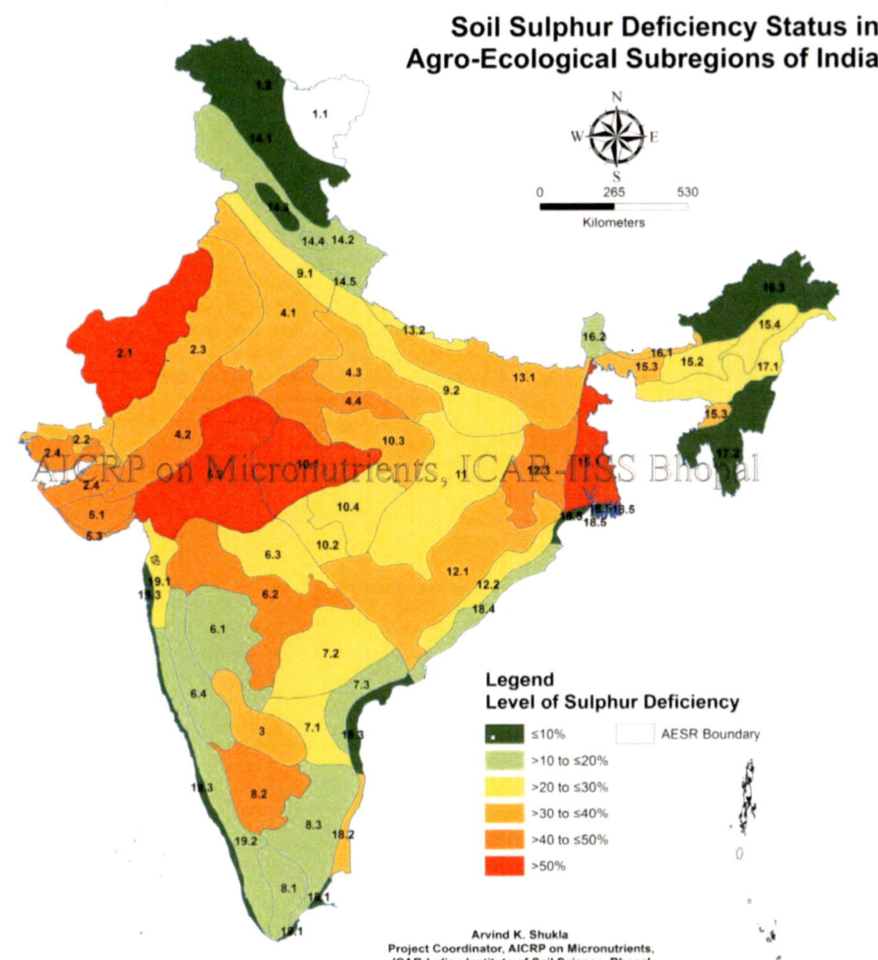

Fig. 10.1. Sulphur deficiency in different agro-ecological zones of India

(with permission from the Project Coordinator, AICRP on Micronutrients, Indian Institute of Soil Science, Bhopal)

Chapter 12: Zinc

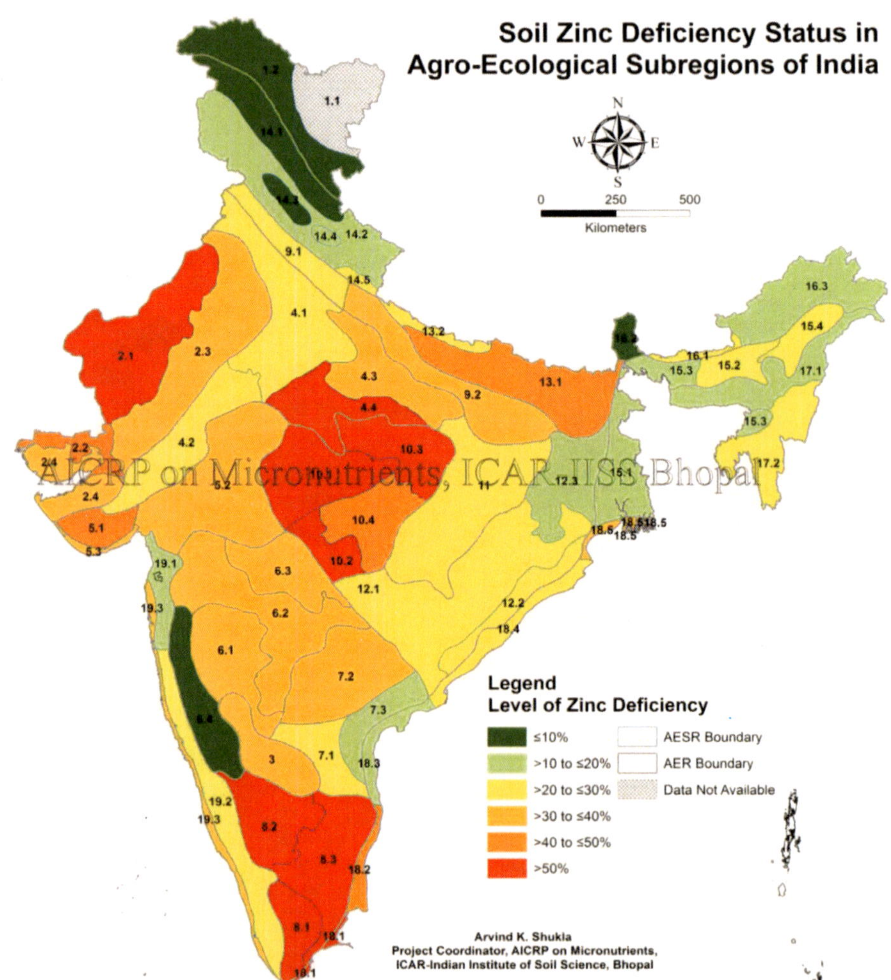

Fig. 12.1: Zn deficiency in different agro-ecological regions of India (with permission from the Project Coordinator, AICRP on Micronutrients, Indian Institute of Soil Science, Bhopal)

Chapter 13: Boron

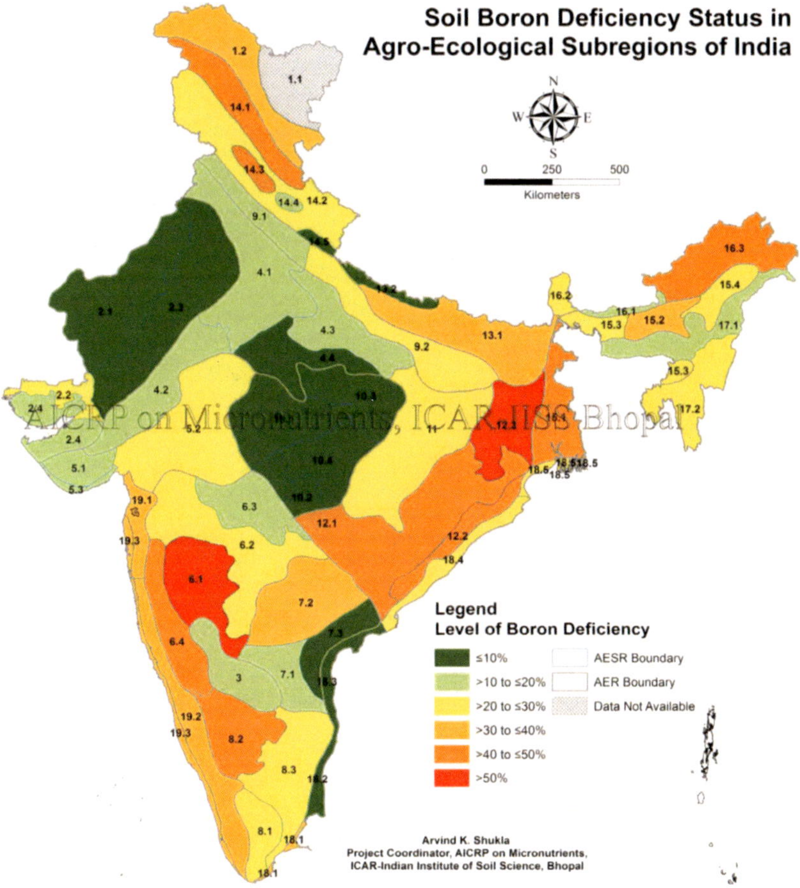

Soil Boron Deficiency Status in
Agro-Ecological Subregions of India
N
W
E
S
0
250
500
Kilometers
AICRP on Micronutrients, ICAR-IISS Bhopal
Legend
Level of Boron Deficiency
≤10%
>10 to ≤20%
>20 to ≤30%
>30 to ≤40%
>40 to ≤50%
>50%
AESR Boundary
AER Boundary
Data Not Available
Arvind K. Shukla
Project Coordinator, AICRP on Micronutrients,
ICAR-Indian Institute of Soil Science, Bhopal

Fig. 13.1: Extent of B deficiency in different agro-ecological zones of India.

(With permission from the Project Coordinator, AICRP on Micronutrients, Indian Institute of Soil Science, Bhopal)

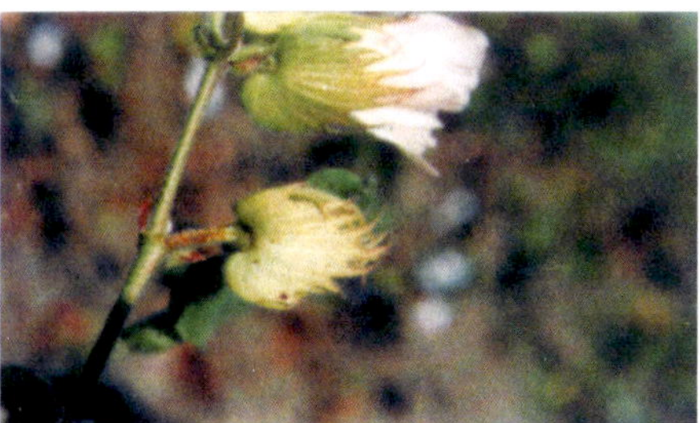

Fig. 13.2: Early stage of abscission of cotton square. Photo courtesy ICAR-Central Institute of Cotton Research, Nagpur, Maharashtra with permission (Perumal *et al*., 2006)

Chapter 14: Organic Manures

Fig.14.1: A specimen (a) of vermicompost made using *Eisenia foetida* (b)

Source: S.S. Rana, Biological Intensive Nutrient Management (BINM)-Vermicompost (via internet)

Fig. 14.2: A biogas plant

Chapter 16: Making Fertilizer Recommendations Based on Soil Tests and Plant Tissue Testing

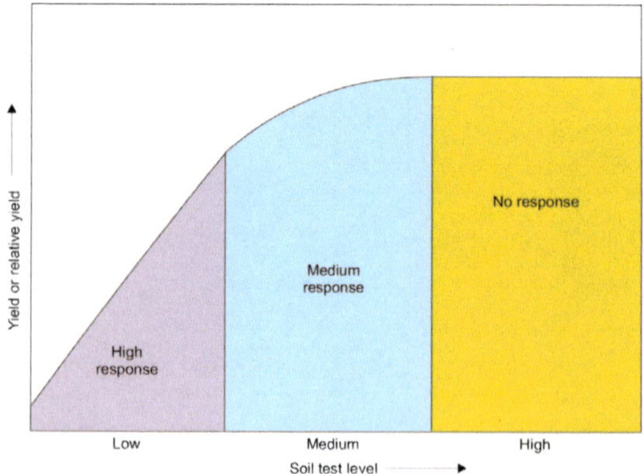

Fig.16.1: Response curve showing relationship between low, medium and high soil-test values and the kind of response expected